# BAHAMAS PRIMARY

# Mathematics Book

The authors and publishers would like to thank the following members of the Teachers' Panel, who have assisted in the planning, content and development of the books, led by Dr Joan Rolle, Senior Education Officer, Primary School Mathematics, Department of Education:

**Deidre Cooper**, Catholic Board of Education

**Vernita Davis**, Ministry of Education Examinations and Assessment Division

**LeAnna T. Deveaux-Miller**, T.G. Glover Professional Development and Research School

**Joelynn Stubbs**, C.W. Sawyer Primary School

**Dyontalee Turnquest Rolle**, Eva Hilton Primary School

Karen
Morrison
and
Lois Lubbe

The Publishers would like to thank the following for permission to reproduce copyright material.

**Photo credits**

**p.37** © Jamie Wilson/123RF.com; **p.60** Bahamian coins © Central Bank of The Bahamas; **p.62** Bahamian bills © Central Bank of The Bahamas; **p.149** © Blend Images/Alamy Stock Photo

**All other photos** © Mike van der Wolk, Tel: +27 83 2686000, mike@springhigh.co.za.

Although every effort has been made to ensure that website addresses are correct at time of going to press, Hodder Education cannot be held responsible for the content of any website mentioned in this book. It is sometimes possible to find a relocated web page by typing in the address of the home page for a website in the URL window of your browser.

Hachette UK's policy is to use papers that are natural, renewable and recyclable products and made from wood grown in sustainable forests. The logging and manufacturing processes are expected to conform to the environmental regulations of the country of origin.

Orders: please contact Bookpoint Ltd, 130 Park Drive, Milton Park, Abingdon, Oxon OX14 4SE. Telephone: (44) 01235 827720. Fax: (44) 01235 400454. Email education@bookpoint.co.uk Lines are open from 9 a.m. to 5 p.m., Monday to Saturday, with a 24-hour message answering service. You can also order through our website: www.hoddereducation.com

ISBN: **978 1 4718 64544**

© Cloud Publishing Services 2016

First published in 2016 by
Hodder Education,
An Hachette UK Company
Carmelite House
50 Victoria Embankment
London EC4Y 0DZ

www.hoddereducation.com

Impression number    10 9 8 7 6 5 4 3 2

Year              2020  2019

Cover photo © kamonrat meunklad/123RF.com

Illustrations by Peter Lubach and Aptara Inc.

Typeset in India by Aptara Inc.

Printed in India

A catalogue record for this title is available from the British Library.

# Contents

# Topic 1 Getting Ready Workbook pages 2–6

**Key Words**
shape
pattern
sets
count
order
before
after
between

▲ What shapes and patterns can you see on this stall? What sets of objects can you see? What can you count?

You learned a lot of mathematics in Grade 1. You need to remember what you learned so that you can build on and learn even more. Before you start learning new things, it is good to look back and remind yourself of the things you already know.

## Getting Started

1 Find three **sets** of objects in the photo. **Count** the objects in each set. Write the amounts in order from smallest to greatest.

2 Choose one **pattern**. Describe it to your partner.

3 If the stall owner sold three hats, how many would she have left?

4 How could the stall owner use a graph to show how many hats, baskets and other items she sold in a week?

# Unit 1 Patterns and Shapes

Shapes, colours, **letters**, numbers, sounds *and* actions *can be repeated to make a* **pattern**.

*You can represent the same pattern in different ways; for example, you can translate the first row of the shape pattern shown to letters or numbers.*

*1 22 1 22    A BB A BB*

1 Translate each pattern into shapes.

   a 1 2 3 1 2 3              b A B A A B A B A A

2 Draw a shape pattern of your own. Ask your partner to translate it into a pattern using letters and a pattern using numbers.

## Looking Back

**a** Which shape has 3 straight sides?

**b** Which shape has no straight sides?

# Unit 2 Counting and Ordering Numbers

## Let's Think ...

- How many baskets are there in this row?
- How can you tell how many there are without counting?
- Describe the basket which comes between basket 7 and basket 9.

1  2  3  4  5  6  7  8  9  10

*You can use a number line to help **count** and **order** numbers.*

**1** Which numbers are missing from this number line?

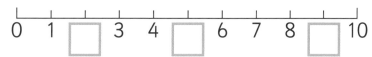

0  1  ☐  3  4  ☐  6  7  8  ☐  10

a How do you know where 2 comes on the number line?

b Why does 7 come **after** 6?

c Why does 9 come **before** 10?

d What number comes **between** 4 and 6?

**2** What is wrong with this number line?

20 19 18 17 16 15 14 13 12 11 10 10 9  8

## Looking Back

Look at the number line and say whether each statement is true or false.

a The ball is at 2.

b The star is at 10.

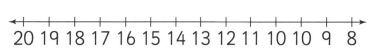

0  1  ●  3  ☐  5  6  7  8  9  10 11  ★  13

c The missing number is smaller than 3.

d The missing number is 1 more than 3.

# Unit 3 Addition and Subtraction

**Let's Think ...**

The total score for rolling two dice is shown.

| Score 7 | Score 8 | Score 7 | Score 10 |

- What is the score on the second dice in each pair?
- How did you work this out?

*You already know the addition and subtraction facts to 10.*
*Addition facts can help you subtract.*
*For example,    6 + 4 = 10   so    10 − 6 = 4   and   10 − 4 = 6*
*If you cannot remember a fact, you can:*

- *count on or back*
- *use counters to model it*
- *make a 10*
- *use a fact you do know to work out the answer.*

1 Add:

   a 3 + 2      b 4 + 4      c 8 + 1      d 5 + 3      e 6 + 4      f 9 + 1

2 Subtract:

   a 7 − 3      b 9 − 2      c 10 − 5      d 8 − 1      e 5 − 5      f 6 − 5

3 Add the numbers on each child's board. Match each child's total with
   the correct fruit basket.

   a                 b                 c                 d                 e

**4** Look at each basket in Question 3. How many more pieces of fruit do you need to make 10 in each basket?

Write a number sentence for each in your copy book.

**5** At a school fair, children throw two stones onto a number board. They win a prize if they score more than 8. Which of these results would win the thrower a prize?

| 7 | −1 |
|---|---|
| ⬤2 | ⬤9 |
| −4 | 8 |

a

| 3 | ⬤−2 |
|---|---|
| −1 | 6 |
| ⬤8 | −4 |

b

| 1 | ⬤7 |
|---|---|
| 8 | ⬤3 |
| −4 | 2 |

c

**6** Sally has 5 beads. Beth has 4 more beads than Sally. How many beads does Beth have?

**7** Micah has 10 marbles. Jayson has 3 fewer. How many marbles does Jayson have?

## Looking Back

**1** Work out the missing number in each number sentence.

**a** 8 + ☐ = 10        **b** 5 + ☐ = 10        **c** 7 + ☐ = 10

**2** The first three digits of Mrs Brown's licence plate are covered in mud.

Use these clues to write the three digits in the correct order.

- The first digit is the answer to 10 − 3.
- The second digit is the answer to double 3.
- The third digit is 1 less than 3.

# Topic Review

## What Did You Learn?

- We revised some of what we learned last year.
- We practiced some number facts.

## Talking Mathematics

1 What is the correct mathematical word for each of these?

a

b

c 2 + 7 = 9

d
| Red | 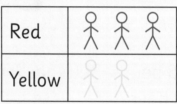 |
| Yellow | |

e ABBABBA

2 Give an example to show what each phrase means.

A circle                    One half                    A subtraction fact

The third item in a row            A doubles fact

## Quick Check

1 One of the number facts does not belong in each set. Which one? Why?

Set A:   5 + 3          4 + 4          9 − 1          2 + 7

Set B:   7 − 4          9 − 6          2 + 2          10 − 7

Set C:   10 − 5         1 + 4          9 − 3          2 + 3

2 Complete each number pattern.

a 2, 4, 4, 2, ☐, ☐    b 1, 2, 2, 1, 2, ☐, ☐    c 5, 10, 5, 5, ☐, ☐

3 Write a number sentence with the answer of 7 using only numbers between 0 and 10. How many different number sentences like this can you make? Write them in a list.

4 Draw a shape with 2 long straight sides and 2 shorter straight sides. Write the name of the shape you have drawn.

# Topic 2  Patterns Workbook pages 7–9

▲ What has been used to make this pattern? How does the pattern work?

We can find patterns everywhere we look: in nature, in our homes, on our school uniforms, in numbers, in shapes and even in games. (Think about how the players are arranged at the start of a game.)

Patterns are very important in mathematics. In this topic, you will learn more about different types of patterns and how you can use them to help you learn and remember facts.

## Getting Started

1 Think about where you could find an example of each of these types of patterns at home or school:

   a patterns with shapes or pictures

   b patterns with sounds or actions

   c patterns in numbers or letters.

2 Share your ideas with your group.

**Key Words**
pattern
repeat
pattern unit
ascending
descending
horizontal
vertical
diagonal

# Unit 1 Describing Patterns

## Let's Think …

- How do you know when something is a pattern?
- Which of these are patterns?
- How do you know?

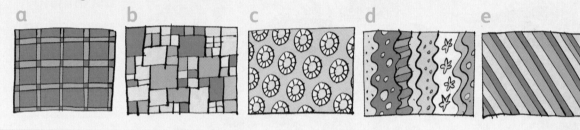

a       b       c       d       e

Some **patterns** have **repeating** units.

The pattern   A   B   A   B   A   B   has two **units**. Each unit is repeated in the same order again and again.

Other patterns do not repeat. Their units may get bigger or smaller in a pattern.

This is a counting pattern of 2.    2, 4, 6, 8, 10 …

The numbers are getting bigger. They are in **ascending** order.

In this counting pattern of 5.     25, 20, 15, 10, 5 …

The numbers get smaller. They are in **descending** order.

1   Which rows form patterns?

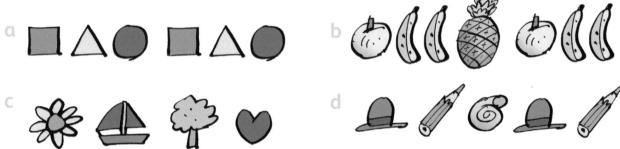

2. Some numbers are missing from this chart. Count from 1 to 30. Say the missing numbers.

| 1 | 2 | 3 | 4 | 5 | 6 | 7 | 8 | | 10 |
|---|---|---|---|---|---|---|---|---|---|
| 11 | | 13 | | | 16 | 17 | | 19 | 20 |
| | 22 | | 24 | | | | 28 | | |

3. Copy and complete these counting patterns. Write *ascending* or *descending* next to each one.

a 5, 6, 7, 8, ☐, ☐, ☐

b 24, 23, 22, 21, ☐, ☐, ☐

c 13, 14, 15, 16, ☐, ☐, ☐

d 18, 19, 20, 21, ☐, ☐, ☐

e 19, 18, 17, 16, ☐, ☐, ☐

*There are lots of patterns in a number chart.*
*Patterns across rows are called* **horizontal** *patterns.*
*Patterns up and down in columns are called* **vertical** *patterns.*

*Patterns that go across the chart are called* **diagonal** *patterns.*

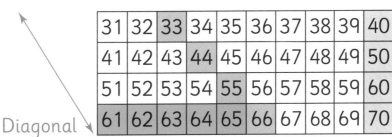

Diagonal

| 31 | 32 | 33 | 34 | 35 | 36 | 37 | 38 | 39 | 40 |
|---|---|---|---|---|---|---|---|---|---|
| 41 | 42 | 43 | 44 | 45 | 46 | 47 | 48 | 49 | 50 |
| 51 | 52 | 53 | 54 | 55 | 56 | 57 | 58 | 59 | 60 |
| 61 | 62 | 63 | 64 | 65 | 66 | 67 | 68 | 69 | 70 |

Vertical

Horizontal

4. How many different number patterns can you make from the number chart? Try to make ascending and descending patterns. Write them in your copy book.

## Looking Back

Janita started at 90 and skip-counted back by 10s until she reached 0. Write down her number pattern.

# Unit 2  Patterns in Number Facts

**Let's Think …**

- Look at these sets of cubes.

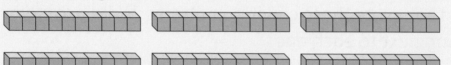

- What pattern can you see?

You can use patterns to work out things you do not remember in mathematics.

The pattern in the cubes above shows all the addition facts for 10.

Look at the numbers in the pattern carefully.

10 + 0 = 10    9 + 1 = 10    8 + 2 = 10    7 + 3 = 10    6 + 4 = 10    5 + 5 = 10

You can also use patterns to work out the fact families for each set of numbers. Look at the fact family for 1, 9 and 10.

9 + 1 = 10    1 + 9 = 10    10 − 1 = 9    10 − 9 = 1

1  Use these cube patterns to write ten addition facts for 11.

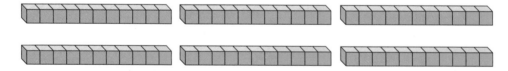

2  Use these counter patterns to write the addition facts for 12. Which fact is not shown here?

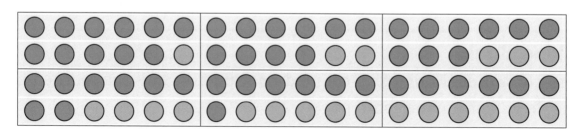

3 Copy and complete these number sentences.

a 10 + ☐ = 11  b 8 + ☐ = 11  c 6 + ☐ = 11

d 10 + ☐ = 12  e 7 + ☐ = 12  f 3 + ☐ = 12

g 11 − 1 = ☐  h 12 − 5 = ☐  i 11 − 8 = ☐

4 True or false?

a 3 + 9 = 12  b 9 + 5 = 12  c 9 + 1 = 11

d 7 + 3 = 11  e 5 + 6 = 11  f 6 + 6 = 12

5 Mrs Plunkett needs 12 buttons. She has 7 buttons. How many more buttons does she need?

6 Jake counted 11 kites in the shop. The next day there were only 4 kites. How many kites had been sold?

**Looking Back**

1 Test your number facts. How quickly can you work out the score on each pair of dice?

a   b   c

d   e   f

2 Write the fact family for 2, 9 and 11.

# Topic Review

## What Did You Learn?

- Some patterns have a repeated design.
- The pattern unit is the part that repeats.
- Some number patterns do not repeat. When the numbers get bigger, the pattern is ascending. When the numbers get smaller, the pattern is descending.
- Patterns can help us to work out and remember number facts.

## Talking Mathematics

Match each word to the correct meaning.

| | |
|---|---|
| pattern | in order from greatest to smallest |
| descending | in a row |
| fact family | appearing over and over |
| horizontal | a set of related additions and subtractions |
| repeating | elements that are arranged in a set way |

## Quick Check

1 Represent the pattern AB AB AB using:

  a shapes

  b colours

  c numbers.

2 Arrange these numbers in ascending order.

  6, 10, 3, 19, 25

3 Arrange these numbers in descending order.

  17, 22, 13, 27, 19

4 Describe this number pattern.

  12, 10, 8, 6, 4, 2

5 11 + 7 = 18

  Write a fact family for 7, 11 and 18.

6 Write the number that is:

  a 1 less than 40

  b 5 less than 35

  c 10 more than 50.

7 A box contains 15 crayons. Misha takes out 7 crayons to colour a picture. How many crayons are left in the box?

# Topic 3 Counting and Number Sense Workbook pages 10–12

▲ Count how many conch are tied together. This fisherman has 7 groups like this. How many is that?

You already know how to count forwards and backwards in ones. You also know that you can use groups and skip numbers to count quickly. This year, you will learn to count to an even higher number than last year. You will also learn more about numbers, including how to decide if they are odd or even.

## Getting Started

1 Count to the highest number that you know.

2 The conch in the photograph are tied in groups of 5. There are 90 conch at the market at Fresh Creek. How many groups of 5 is this? How did you work out the answer?

3 Michelle is thinking of a number. It has a 7 in the ones place. What number will be in the ones place in the next number? Does it matter which numbers are in the other places?

# Unit 1 Skip-Counting

**Let's Think ...**

● How would you count to find the total for each group?

● Explain why.

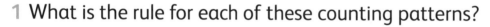

*Remember we use* **skip-counting** *when we want to count* **groups** *of objects quickly.*

● *We count groups of two by 2s.*    *0, 2, 4, 6, 8, 10...*

● *We count groups of five by 5s.*    *0, 5, 10, 15, 20...*

● *We count groups of ten by 10s.*    *0, 10, 20, 30, 40...*

1 What is the rule for each of these counting patterns?

   a 2, 4, 6, 8, 10, 12, 14, 16, 18, 20

   b 10, 20, 30, 40, 50, 60, 70, 80, 90, 100

   c 5, 10, 15, 20, 25, 30, 35, 40, 45, 50

   d 12, 10, 8, 6, 4, 2

   e 30, 25, 20, 15, 10, 5

   f 80, 70, 60, 50, 40, 30, 20, 10

2 Which number does not belong in each skip-counting pattern? How do you know?

   a 2, 4, 6, 8, 9, 10                    b 22, 20, 18, 16, 14, 13, 10

   c 5, 10, 15, 20, 23, 30              d 100, 90, 80, 70, 60, 65, 50

| 1 | 2 | 3 | 4 | 5 | 6 | 7 | 8 | 9 | 10 |
|---|---|---|---|---|---|---|---|---|---|
| 11 | 12 | 13 | 14 | 15 | 16 | 17 | 18 | 19 | 20 |
| 21 | 22 | 23 | 24 | 25 | 26 | 27 | 28 | 29 | 30 |
| 31 | 32 | 33 | 34 | 35 | 36 | 37 | 38 | 39 | 40 |
| 41 | 42 | 43 | 44 | 45 | 46 | 47 | 48 | 49 | 50 |
| 51 | 52 | 53 | 54 | 55 | 56 | 57 | 58 | 59 | 60 |
| 61 | 62 | 63 | 64 | 65 | 66 | 67 | 68 | 69 | 70 |
| 71 | 72 | 73 | 74 | 75 | 76 | 77 | 78 | 79 | 80 |
| 81 | 82 | 83 | 84 | 85 | 86 | 87 | 88 | 89 | 90 |
| 91 | 92 | 93 | 94 | 95 | 96 | 97 | 98 | 99 | 100 |

3 Use the chart for this question.

    a Skip-count by 2s from 14 to 100.    b Skip-count by 5s from 25 to 80.

    c Skip-count back by 2s from 60.    d Skip-count back by 10s from 90 to 40.

    e Skip-count by 10s from 45 to 95.

4 True or false?

    a 46 is 2 **less than** 48.    b 55 is 10 more than 45.

    c 55 is 10 **more than** 65.    d 90 is 5 less than 85.

    e When you count by 2s from 12 to 24, you will count the number 21.    f 73 is 10 more than 63.

## Looking Back

Write the next two numbers in each counting pattern.

**a** 12, 14, 16, ☐, ☐    **b** 46, 56, 66, 76, ☐, ☐

**c** 55, 50, 45, 40, ☐, ☐    **d** 87, 77, 67, ☐, ☐

**e** 31, 33, 35, 37, ☐, ☐

# Unit 2 Counting Above 100

## Let's Think …

- Zachary has 100 chickens. He buys 25 more at the market. How many does he have now?
- Maria has 100 stickers. Her twin sister also has 100 stickers. How many do they have altogether?
- How did you decide each time? Tell your group.

You already know how to count to 100 and that numbers work in a pattern as you count higher and higher.

Listen carefully to the number names and follow on the chart as your teacher counts from one hundred one (101) to two hundred (200).

| 101 | 102 | 103 | 104 | 105 | 106 | 107 | 108 | 109 | 110 |
|-----|-----|-----|-----|-----|-----|-----|-----|-----|-----|
| 111 | 112 | 113 | 114 | 115 | 116 | 117 | 118 | 119 | 120 |
| 121 | 122 | 123 | 124 | 125 | 126 | 127 | 128 | 129 | 130 |
| 131 | 132 | 133 | 134 | 135 | 136 | 137 | 138 | 139 | 140 |
| 141 | 142 | 143 | 144 | 145 | 146 | 147 | 148 | 149 | 150 |
| 151 | 152 | 153 | 154 | 155 | 156 | 157 | 158 | 159 | 160 |
| 161 | 162 | 163 | 164 | 165 | 166 | 167 | 168 | 169 | 170 |
| 171 | 172 | 173 | 174 | 175 | 176 | 177 | 178 | 179 | 180 |
| 181 | 182 | 183 | 184 | 185 | 186 | 187 | 188 | 189 | 190 |
| 191 | 192 | 193 | 194 | 195 | 196 | 197 | 198 | 199 | 200 |

Say the next few numbers using the same counting pattern.

1 Say the next three numbers in each counting pattern.

a 182, 183, 184, …    b 115, 116, 117, …    c 120, 119, 118, …

d 187, 186, 185, …    e 157, 158, 159, …

**2** Say the number that comes after:

a 126　　　　b 187　　　　c 199　　　　d 200

**3** Say the number that comes before:

a 145　　　　b 188　　　　c 120　　　　d 150

**4** What number comes **between**:

a 139 and 141　　b 199 and 201?

**5** Read the number names. Write the numerals.

a one hundred twenty-five　　b one hundred forty

c one hundred ninety-nine　　d two hundred

**6** Which number is smaller in each pair? Say the number name.

a 109 or 190　　b 127 or 172　　c 189 or 198

**7** Write each set of numbers in order from smallest to greatest.

a 187, 168, 178

b 145, 104, 115

c 163, 136, 166

**8** True or false?

197 is less than 179.

201 is more than 200.

345 is more than 300.

299 is less than 300.

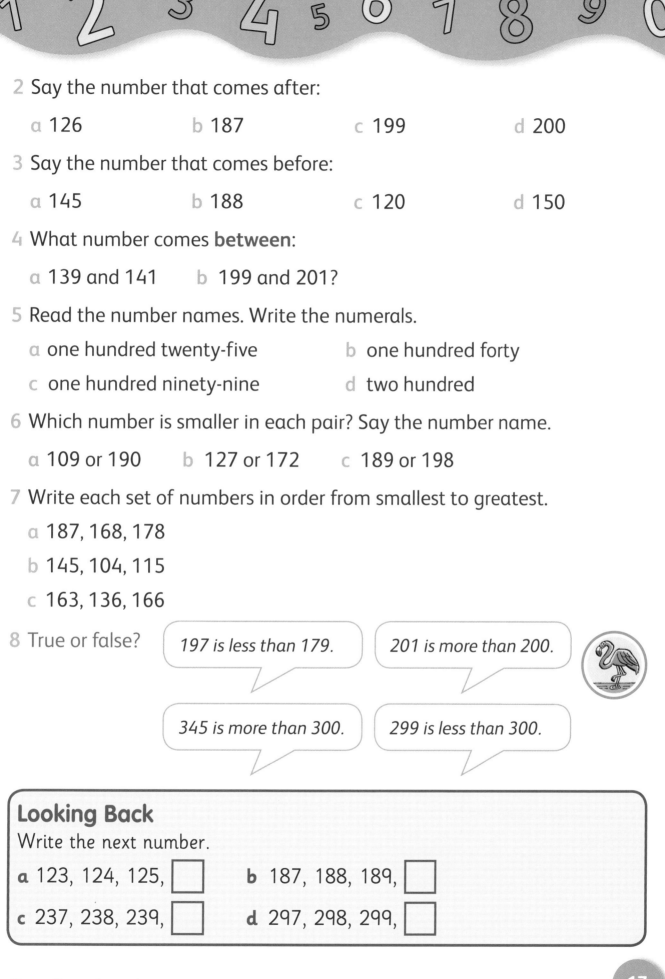

**Looking Back**

Write the next number.

**a** 123, 124, 125, ☐　　**b** 187, 188, 189, ☐

**c** 237, 238, 239, ☐　　**d** 297, 298, 299, ☐

# Unit 3  Odd and Even Numbers

**Let's Think …**

- How many counters are in each set?
- Which sets can be arranged in pairs?
- Which sets have one counter left over?

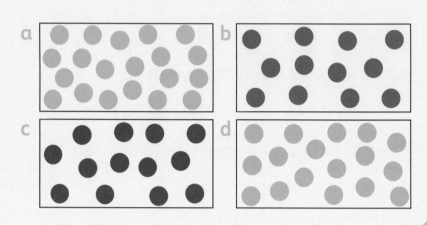

a    b

c    d

---

Numbers that can be arranged in pairs with none left over are called **even** numbers.

Numbers that have one left over when you arrange them in pairs are called **odd** numbers.

▲ 4 and 6 are even numbers.    ▲ 3 and 5 are odd numbers.

This is the pattern of even numbers. Even numbers all end in 0, 2, 4, 6 or 8.

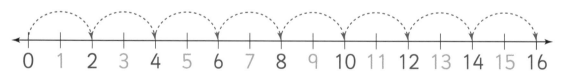

0  1  2  3  4  5  6  7  8  9  10  11  12  13  14  15  16

This is the pattern of odd numbers. Odd numbers all end in 1, 3, 5, 7 or 9.

0  1  2  3  4  5  6  7  8  9  10  11  12  13  14  15  16

1 Look at these sets of dots.

  a Which sets are odd? Say the numbers.

  b Which sets are even? Say the numbers.

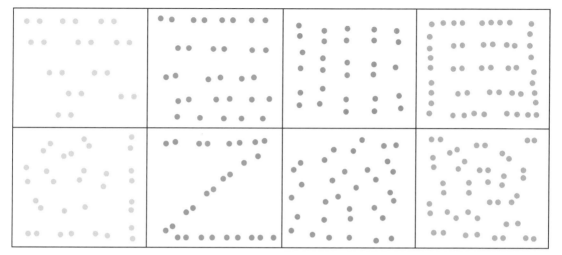

2 Say which numbers are even and which numbers are odd.

  a 13          b 26          c 88          d 100

  e 117         f 225         g 289         h 202

3 Look at the house numbers in this street.

  a How are the houses numbered?

  b What are the missing house numbers?

  c What is the next house number on each side of the street? How do you know?

UPALONG STREET

26  28  30  32  34  ?  38  ?

27  29  31  33  ?  37  39  ?

---

**Looking Back**

**1** Write down all the odd numbers in this set.
   22, 32, 43, 21, 19, 45, 64, 100, 99, 48

**2** Write down five even numbers between 120 and 140.

# Topic Review

**Talking Mathematics**

Complete these sentences.

a The number after 134 is …

b We get the numbers 35, 40, 45, 50 when we …

c Numbers that end with 1, 3 or 5 are called …

d I know a number is an even number if …

**Quick Check**

1 Write the numbers that complete these sentences.

  a $11 + 3 =$ ☐

  b $19 - 1 =$ ☐

  c $40 + 5 =$ ☐

  d Double 5 = ☐

  e The even number after 42 is ☐.

  f The odd number before 17 is ☐.

2 Write the number that comes between:

  a 87 and 89    b 120 and 122    c 189 and 191

3 Use the clues to work out the missing numbers.

  ☐ , ☐ , 84, ☐ , ☐

  - The numbers are in ascending order.
  - The biggest number is 10 more than 100.
  - The third missing number is the odd number after 91.
  - The smallest number is 10 less than 60.
  - The second number is even. It is bigger than 64, but smaller than 68.

# Topic 4  Talking about Time

Workbook pages 13–15

**Bahamas Junkanoo Carnival**
3 marvellous days of fun in MAY each year

## Key Words
day
week
month
year
calendar
date
leap year

▲ Can you find three time words in this advert?

Time is an important part of daily life. In this topic, you will revise some of the words you learned last year to talk about time. You will learn to use calendars and solve problems that involve different units of time.

## Getting Started

1 Look at the list of key words. Talk in groups about what each word means.

2 Why do people use calendars? Think of as many things as possible. Share your ideas with your group.

3 Find today's date on the class calendar.

4 What will the date be tomorrow?

5 What was the date two days ago?

6 How many days are there in June?

# Unit 1  Time to Remember

**Let's Think ...**

● How many days are there in a week?

● Say the names of the days of the week in order.

● What month is it now?

● How many more months are there until the end of the year?

A **week** *is the same amount of time as seven* **days**.

*Sunday   Monday   Tuesday   Wednesday   Thursday   Friday   Saturday*

A **year** *is the same amount of time as 12 months.*

*January   February   March   April   May   June*
*July   August   September   October   November   December*

One **month** *is about 4 weeks.*

1 Write the answers.

   a A week has ⎯ days.

   b Saturday and ⎯ are the days of the weekend.

   c The day after Monday is ⎯.

   d The day after Thursday is ⎯.

   e The day before Wednesday is ⎯.

   f The day before Monday is ⎯.

2 Use the year planner on page 23 to answer these questions.

   a Which month comes after March?

   b Which month comes after June?

   c In which month did you start school?

   d Which month comes before February?

   e Which month comes between August and October?

   f Which month is the last month of the year?

## YEAR PLANNER

| January | February | March | April |
|---|---|---|---|
| May *Junkanoo | June *Eleuthera Pineapple Fest | July | August |
| September | October | November | December *Christmas |

3 Some special events are marked on the planner.

  a What are they?

  b In which month does each event take place?

4 Work in groups.

  • Make your own planner for the year.

  • Write your name and birthday in the correct place on the planner.

  • Mark three other important events on the planner.

### Looking Back

Read this poster.

a When is closed season for grouper?

b For how many months are fishermen not allowed to catch grouper?

c Is it legal to fish for grouper in January?

Nassau Grouper Season will be **CLOSED** December 1st – February 28th Every Year

# Unit 2  Working with a Calendar

A **calendar** *shows how the year is broken up into months, weeks and days.*

*The days of the month are numbered. You can look at a calendar to find out what day of the week it is on any given* **date***.*

**1 a** Which months have 31 days?

**b** Which month is the sixth month?

**c** Which month has less than 30 days?

**d** Is September the ninth month?

**e** How many weekends are there in August?

**f** How many weeks are there in May?

**g** Which months have 30 days?

**h** How many Mondays are there in June?

**i** The last month of the year is December. What is the ordinal number for this?

**2** Find the month it is now on a calendar. What day of the week is each date?

**a** 2nd     **b** 9th     **c** 11th     **d** 20th

**3** Use a calendar to solve these problems.

**a** Keisha's birthday month is before August, but after June. The date of her birthday is double 6. When is her birthday?

**b** On which day of the week is her birthday this year?

**c** How long is it until Keisha's next birthday?

## Looking Back

On what day of the week are the following days on this calendar?

**a** Majority Rule Day (10 January)

**b** Labor Day (7 June)

**c** Independence Day (10 July)

**d** Emancipation Day (5 August)

| JANUARY | | | | | | |
|---|---|---|---|---|---|---|
| M | T | W | T | F | S | S |
| | 1 | 2 | 3 | 4 | 5 | |
| 6 | 7 | 8 | 9 | 10 | 11 | 12 |
| 13 | 14 | 15 | 16 | 17 | 18 | 19 |
| 20 | 21 | 22 | 23 | 24 | 25 | 26 |
| 27 | 28 | 29 | 30 | 31 | | |

| FEBRUARY | | | | | | |
|---|---|---|---|---|---|---|
| M | T | W | T | F | S | S |
| | | | | | 1 | 2 |
| 3 | 4 | 5 | 6 | 7 | 8 | 9 |
| 10 | 11 | 12 | 13 | 14 | 15 | 16 |
| 17 | 18 | 19 | 20 | 21 | 22 | 23 |
| 24 | 25 | 26 | 27 | 28 | | |

| MARCH | | | | | | |
|---|---|---|---|---|---|---|
| M | T | W | T | F | S | S |
| | | | | | 1 | 2 |
| 3 | 4 | 5 | 6 | 7 | 8 | 9 |
| 10 | 11 | 12 | 13 | 14 | 15 | 16 |
| 17 | 18 | 19 | 20 | 21 | 22 | 23 |
| 24 | 25 | 26 | 27 | 28 | 29 | 30 |
| 31 | | | | | | |

| APRIL | | | | | | |
|---|---|---|---|---|---|---|
| M | T | W | T | F | S | S |
| | 1 | 2 | 3 | 4 | 5 | 6 |
| 7 | 8 | 9 | 10 | 11 | 12 | 13 |
| 14 | 15 | 16 | 17 | 18 | 19 | 20 |
| 21 | 22 | 23 | 24 | 25 | 26 | 27 |
| 28 | 29 | 30 | | | | |

| MAY | | | | | | |
|---|---|---|---|---|---|---|
| M | T | W | T | F | S | S |
| | | 1 | 2 | 3 | 4 | |
| 5 | 6 | 7 | 8 | 9 | 10 | 11 |
| 12 | 13 | 14 | 15 | 16 | 17 | 18 |
| 19 | 20 | 21 | 22 | 23 | 24 | 25 |
| 26 | 27 | 28 | 29 | 30 | 31 | |

| JUNE | | | | | | |
|---|---|---|---|---|---|---|
| M | T | W | T | F | S | S |
| | | | | | | 1 |
| 2 | 3 | 4 | 5 | 6 | 7 | 8 |
| 9 | 10 | 11 | 12 | 13 | 14 | 15 |
| 16 | 17 | 18 | 19 | 20 | 21 | 22 |
| 23 | 24 | 25 | 26 | 27 | 28 | 29 |
| 30 | | | | | | |

| JULY | | | | | | |
|---|---|---|---|---|---|---|
| M | T | W | T | F | S | S |
| | 1 | 2 | 3 | 4 | 5 | 6 |
| 7 | 8 | 9 | 10 | 11 | 12 | 13 |
| 14 | 15 | 16 | 17 | 18 | 19 | 20 |
| 21 | 22 | 23 | 24 | 25 | 26 | 27 |
| 28 | 29 | 30 | 31 | | | |

| AUGUST | | | | | | |
|---|---|---|---|---|---|---|
| M | T | W | T | F | S | S |
| | | | | 1 | 2 | 3 |
| 4 | 5 | 6 | 7 | 8 | 9 | 10 |
| 11 | 12 | 13 | 14 | 15 | 16 | 17 |
| 18 | 19 | 20 | 21 | 22 | 23 | 24 |
| 25 | 26 | 27 | 28 | 29 | 30 | 31 |

| SEPTEMBER | | | | | | |
|---|---|---|---|---|---|---|
| M | T | W | T | F | S | S |
| 1 | 2 | 3 | 4 | 5 | 6 | 7 |
| 8 | 9 | 10 | 11 | 12 | 13 | 14 |
| 15 | 16 | 17 | 18 | 19 | 20 | 21 |
| 22 | 23 | 24 | 25 | 26 | 27 | 28 |
| 29 | 30 | | | | | |

| OCTOBER | | | | | | |
|---|---|---|---|---|---|---|
| M | T | W | T | F | S | S |
| | 1 | 2 | 3 | 4 | 5 | |
| 6 | 7 | 8 | 9 | 10 | 11 | 12 |
| 13 | 14 | 15 | 16 | 17 | 18 | 19 |
| 20 | 21 | 22 | 23 | 24 | 25 | 26 |
| 27 | 28 | 29 | 30 | 31 | | |

| NOVEMBER | | | | | | |
|---|---|---|---|---|---|---|
| M | T | W | T | F | S | S |
| | | | | | 1 | 2 |
| 3 | 4 | 5 | 6 | 7 | 8 | 9 |
| 10 | 11 | 12 | 13 | 14 | 15 | 16 |
| 17 | 18 | 19 | 20 | 21 | 22 | 23 |
| 24 | 25 | 26 | 27 | 28 | 29 | 30 |

| DECEMBER | | | | | | |
|---|---|---|---|---|---|---|
| M | T | W | T | F | S | S |
| 1 | 2 | 3 | 4 | 5 | 6 | 7 |
| 8 | 9 | 10 | 11 | 12 | 13 | 14 |
| 15 | 16 | 17 | 18 | 19 | 20 | 21 |
| 22 | 23 | 24 | 25 | 26 | 27 | 28 |
| 29 | 30 | 31 | | | | |

# Unit 3  Equivalent Units of Time

**Let's Think …**

Mr Dorsett says he will take 15 days to paint a building.

Mrs Pinder says she will take two weeks.

● Who will take longer?

● How did you decide?

There are seven days in a week.    7 days = 1 week

There are 365 days in a year.    365 days = 1 year

Every fourth year has 366 days. This is called a **leap year**. In a leap year, there is an extra day added, 29 February.

There are 12 months in a year.    12 months = 1 year

Your teacher will give you a calendar for this year to complete these activities.

1 Is this year a leap year? How do you know?

2 How many years is it until the next leap year?

3 Find today's date. How many weeks are left in this month?

4 How many full weeks are there:

   a in August

   b from the start of June to the end of July

   c in the year?

**5** How many weeks is it until Christmas? Is this more, or less, than 40 days?

**6** How many months pass from 1 January to 30 June?

**7** How many days in this month will you spend at school?

**8** Zion says there are 28 days in 4 weeks. Is he correct?

**9** Arrange these times in order, from the shortest to the longest.

| | | | |
|---|---|---|---|
| 30 weeks | 6 months | 17 days | 3 weeks |
| 1 leap year | 365 days | 4 years | 13 months |
| 85 days | 100 weeks | | |

**10** How many years have you been alive? How many months is this?

**11** Match the amounts of time that are the same.

| 7 days |
|---|
| 12 months |
| 1 day |
| 366 days |

| 1 year |
|---|
| 24 hours |
| 1 leap year |
| 1 week |

---

**Looking Back**

Copy and complete these statements.

**a** 24 hours = ____

**b** 7 days = ____

**c** 12 months = ____

**d** 365 days = ____

**e** 366 days = ____

# Topic Review

## What Did You Learn?

- We revised the names of days and months.
- A calendar shows the dates, days and months of the year.
- Different units of time are equivalent to each other.
  24 hours = 1 day      7 days = 1 week      12 months = 1 year
- There are 365 days in a year unless it is a leap year.

## Talking Mathematics

Write down one word that means the same as each phrase.

a period of 12 months      the third day of the week      24 hours

7 days      the tenth month      a diagram showing the days, dates and months

## Quick Check

1 a  Which month is this calendar page for?

  b  How many Thursdays are there in this month?

  c  What day is the 9th of this month?

  d  What month comes before this month?

  e  Find the red star on the calendar. What date would it be on this day in a leap year?

  f  How many full weeks are there in this month?

2  Arrange these number cards to make an odd number between 100 and 200.

3  Jessica added two odd numbers together. Her answer was 14. What numbers could she have added?

# Topic 5 More Counting and Number Facts
Workbook pages 16–18

**Key Words**
count
forwards
backwards
double
near double
add
subtract
fact family

▲ Do you think there are more than 100 beads in this bag? Are there more than 200? About how many beads are in the bag? How could you be sure?

You have used number patterns to count from 0 to 200. You will now count even higher numbers. You will also practice the number facts you learned last year and learn some new ones.

## Getting Started

1 Amira buys some beads in packets of 100. How can she count the beads?

2 Amira has two packets of 100 beads and 23 loose beads. How many beads is this?

3 How many beads are there in five packets of 100 beads?

# Unit 1 Counting

**Let's Think …**

Jessica says that a number that ends with 9 in the ones place is always bigger than a number that ends with 6 in the ones place.

Sandra says that this is not true. She says that some numbers with 6 in the ones place are bigger than numbers with 9 in the ones place.

● Which girl is correct? Explain why and give examples.

When you **count forwards** by ones, each number is 1 more than the number before it.

This pattern applies to all numbers, however high you count.

**1** Look carefully at this number chart.

| 121 | 122 | 123 | 124 | 125 | 126 | 127 | a | 129 | |
|-----|-----|-----|-----|-----|-----|-----|-----|-----|-----|
| 131 | b | 133 | | | | | | 139 | |
| c | 142 | 143 | | | 146 | d | | | |
| 151 | 152 | | | | | 157 | | 159 | |
| | 162 | 163 | | e | 166 | | | | |
| 171 | | 173 | | | | | 178 | f | |
| 181 | 182 | 183 | 184 | | | g | | 189 | |
| 191 | 192 | 193 | h | | 196 | | 198 | | 200 |

a Work out which numbers the letters a to h represent.

b Start at 125 and count down the orange column. What is the pattern?

c Start at 200 and count up the green column. What is the pattern?

**Looking Back**

What is the next number after 200? How do you know?

# Unit 2 Doubles Facts

**Let's Think …**

Look at the dice and the domino.

- What do you notice?
- What is the total score on the dice?
- How many dots are on the domino?
- What do we call adding two numbers that are the same?

*We use the word **doubles** to talk about adding two numbers that are the same.*

*3 + 3 = 6          Double 3 is 6*

*You need to learn the doubles of numbers from 1 to 10.*

1 What is:

a double 2          b double 4          c double 7?

2 What is the cost of two of each of these items?

a
$1.00

b
$3.00

c
$7.00

d
$9.00

3 Destiny has $10.00. Her grandpa gives her $10.00 for her birthday. How much does she have now?

4 Mark has 6 marbles. Mike has double that number.

a How many marbles does Mike have?

b How many marbles do they have in all?

5 Marie threaded 9 beads onto a string. Lucy threaded twice as many as Marie. How many did Lucy thread?

## Nearly Double

> *Once you know your doubles, you can use them to add and subtract numbers that are nearly double.*
>
> *For example:*
>
> *2 + 2 = 4*     *2 + 3 is one more than 2 + 2*   *so*   *2 + 3 = 4 + 1 = 5*
>
> *10 + 11 is one more than double 10*
> *Double 10 is 20 so 10 + 11 is double 10 plus one more = 21*
>
> *3 + 4 is one less than double 4*
> *Double 4 is 8 so 3 + 4 is double 4 less one = 7*

1   How could you use doubles to work out the answer to each addition? Write down your method.

    a  5 + 6 = ☐        b  7 + 8 = ☐        c  6 + 7 = ☐

    d  8 + 10 = ☐      e  9 + 10 = ☐      f  10 + 12 = ☐

2   What is:

    a  double 4 plus 2       b  double 6 less 1       c  double 9 plus 2

    d  double 8 less 1       e  double 6 plus 1       f  double 5 less 1?

3   The total for two cards is shown. Work out the number on the card that is face down in each pair.

   a          b          c          d

   Total: 8      Total: 11      Total: 12      Total: 17

### Looking Back

Matt has 9 stickers. His friend James has 1 more than him.
How many do they have altogether?

# Unit 3 Number Facts

**Let's Think …**

- Here is one way to show 13 using red and yellow counters.

○ ● ● ● ● ● ● ● ● ● ● ● ●

1 + 12 = 13

- In what other ways can you show 13 using red and yellow counters? Write them all down.

You can use counters to model number facts.

You can also use your fingers or a number line to help you.

1 Work out the answers. Use a number line if you need to. What is:

    a 2 more than 12       b 2 more than 16       c 3 more than 11

    d 4 more than 14       e 5 more than 10       f 6 more than 13

    g 3 more than 17       h 4 more than 9        i 2 more than 18?

2 One domino in each set shows a total of 14. Which one is it?

    a                                    b

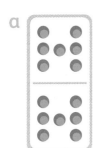   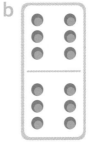 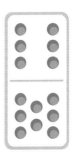 

## Fact Families

A **fact family** *is a set of number sentences that gives the addition and subtraction facts for the same three numbers.*

*You can use one fact to work out the other three facts in a family. The numbers 3, 4 and 7 make this fact family:*

*4 + 3 = 7    3 + 4 = 7    7 − 3 = 4    7 − 4 = 3*

3 Use the first facts to work out the missing numbers in the second facts.

a 14 + 1 = 15    15 − 1 = ☐    b 10 + 5 = 15    15 − 5 = ☐

c 7 + 8 = 15    15 − 8 = ☐    d 15 − 3 = 12    3 + ☐ = 15

e 15 − 6 = 9    6 + ☐ = 15    f 15 − 2 = ☐    2 + ☐ = 15

g 15 − 8 = ☐    8 + ☐ = 15    h 4 + ☐ = 15    15 − 4 = ☐

i 9 + ☐ = 15    15 − 9 = ☐    j 0 + 15 = ☐    15 − 15 = ☐

4 Write the fact family for each set of numbers. Write them in your copy book.

a 3, 5, 8          b 8, 5, 13          c 8, 3, 11

d 7, 4, 11         e 9, 5, 14          f 5, 10, 15

5 You are told that 12 + 9 = 21. What else can you work out?

### Looking Back

Work out the total on each set of cards.

a 12 7    b 16 4    c 19 1    d 12 8    e 8 9

# Unit 4 Number Check

**Let's Think …**

● How do you work out what the missing numbers are in a number chart? Tell your partner.

1 Count aloud from 191 to 250. Say all the missing numbers as you count.

| | | | | | | | | | | |
|---|---|---|---|---|---|---|---|---|---|---|
| row 1 | 191 | 192 | 193 | 194 | 195 | 196 | 197 | 198 | 199 | 200 |
| row 2 | 201 | 202 | 203 | 204 | 205 | 206 | 207 | 208 | 209 | 210 |
| row 3 | 211 | | 213 | | 215 | | 217 | | 219 | |
| row 4 | 221 | 222 | 223 | 224 | 225 | 226 | 227 | 228 | 229 | 230 |
| row 5 | | 232 | | 234 | | 236 | | 238 | | 240 |
| row 6 | 241 | 242 | 243 | 244 | 245 | 246 | 247 | 248 | 249 | 250 |

2 Skip-count by 2s from 202 to 250.

3 Skip-count back by 5s from 250 to 200.

4 Use the chart to answer these questions. Use ordinal numbers to describe the row.

   a Which row has even numbers missing?

   b Which row has odd numbers missing?

   c Which row ends with 210?

   d In which row is the number 226?

   e What number is 10 more than 230? In which row is this number?

   f Which row contains the number that is 10 less than 250?

**Looking Back**

Are these numbers in ascending order?

a 193, 194, 192, 195, 196          b 239, 240, 241, 242, 245

# Topic Review

**Talking Mathematics**

- Is 234 bigger or smaller than 243?
- How do you know whether one number is bigger than another? Explain to your partner.

**Quick Check**

1 Write the missing numbers.

   a 197, 198, 199, ☐, ☐, ☐          b 205, 204, 203, ☐, ☐, ☐

   c 246, 247, ☐, 249, ☐, ☐

2 Write the number that comes after:

   a 145          b 222          c 250          d 300

3 Write the number that comes before:

   a 206          b 330          c 201          d 300

4 What number comes between:

   a 239 and 241          b 299 and 301?

5 Write the other three facts in this family.

   14 + 5 = 19

6 Match these numbers in pairs to make 17.

   6          15          11          2          4          13

# Topic 6  Temperature  Workbook pages 19–20

**Key Words**
temperature
hot
cold
warm
cool
higher
lower
thermometer

▲ Shanay is feeling sick. What has her mom done to see whether she has a fever or not?

Last year you learned about **hot** and **cold** things. Iced water and air-conditioned rooms are cold. Cups of tea and sunny places are hot. **Temperature** is a measure of how hot or cold things are. In this topic, you will learn more about measuring and comparing temperatures.

## Getting Started

1 How many cold things can you think of? Make a list.

2 Name five things that are hot. Which of them is the hottest?

3 When someone has a fever, we say they have a high temperature. What does this mean?

# Unit 1 Comparing Temperatures

## Let's Think ...

Choose the correct word to complete each sentence. Tell your partner why you chose each word.

- Ice is (*colder/warmer*) than the water in the sea.
- It is (*cooler/warmer*) in the early morning than in the afternoon.
- Summer is (*hotter/colder*) than winter.

We can use the words in the diagram to compare **temperatures**.

- *Ice is* **colder** *than water.*
- *It is* **warmer** *in the sun than in the shade.*

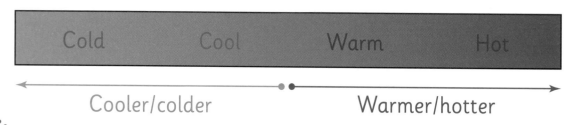

| Cold | Cool | Warm | Hot |

← Cooler/colder        Warmer/hotter →

**1** Look at the pictures. Say which is hot and which is cold.

a

b

c

d

**2** Which is hotter? How do you know?

**3** Which is cooler? How do you know?

**4** Write the items in order, from hottest to coldest.

**5 a** Think about the summer. Describe the temperature when you wake up in the morning. Compare it to the temperature in the evening.

**b** How do you know when the temperature is cooler?

**c** In summer, which part of the day is the hottest?

**d** Compare the temperature of the sea in the morning and at noon. If it is not the same, say why.

---

**Looking Back**

**1** Why is ice-cream kept in a freezer?

**2** Why do you enjoy cold water in summer and a warm drink in winter?

# Unit 2  Using a Thermometer

**Let's Think …**

Mr Johnstone is complaining about the heat. He says 'It feels like 200 degrees out there!'

What do you think he means?

You use a **thermometer** to measure temperature.

The scale on the thermometer is marked in degrees Fahrenheit.

The temperature makes the liquid inside the thermometer move up and down the scale.

You read the temperature by looking at where the top of the line of liquid is on the scale.

This thermometer shows a temperature of 90 degrees Fahrenheit.

When something is cold, it has a **low** temperature.

When something is hot, it has a **high** temperature.

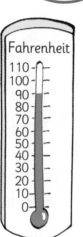

1  What temperatures are shown on these thermometers?

a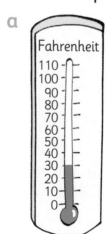

b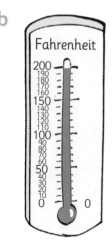

c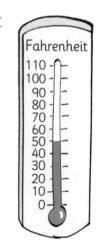

2 Water freezes and turns to ice at 32 degrees Fahrenheit. Water boils at 212 degrees Farhenheit. Look at the thermometers in Question 1.

  a Which thermometer has been inside a freezer?

  b Which thermometer has been inside a hot oven?

  c Where do you think the other thermometer might have been?

3 Miss Rose is a teacher on Spanish Wells. She made this temperature chart for the classroom.

| Month | Jan | Feb | Mar | Apr | May | June | July | Aug | Sep | Oct | Nov | Dec |
|---|---|---|---|---|---|---|---|---|---|---|---|---|
| Temperature (degrees Fahrenheit) | 75 | 78 | 80 | 82 | 85 | 86 | 88 | 90 | 88 | 85 | 82 | 78 |

  a What does the chart show?

  b Which month was hottest?

  c Which month was coldest?

  d In which months was the temperature 85 degrees Fahrenheit?

  e What temperature did she record for September?

**Looking Back**

The temperature does not stay the same all day. Draw a line to show what happens to the temperature through the day. Your line should go up when the temperature gets hotter and down when the temperature gets cooler. Start with a dot to show first thing in the morning.

# Topic Review

**Talking Mathematics**

Draw pictures to show what these words mean.

very hot     boiling     freezing     cool     very cold     thermometer

**Quick Check**

1 Look at the temperatures on these thermometers.

A  Fahrenheit
110
100
90
80
70
60
50
40
30
20
10
0

B  Fahrenheit
110
100
90
80
70
60
50
40
30
20
10
0

C  Fahrenheit
200
190
180
170
160
150
140
130
120
110
100
90
80
70
60
50
40
30
20
10
0

D  Fahrenheit
200
190
180
170
160
150
140
130
120
110
100
90
80
70
60
50
40
30
20
10
0

a Which thermometer shows the coldest temperature?

b Which thermometer shows the temperature of a normal day in summer?

c Which thermometer shows the temperature of a pot of food boiling?

2 Say or write these temperatures in descending order.

109°F, 201°F, 145°F, 154°F, 210°F, 190°F

3 11 + 12 = 23    What is 23 − 11?

4 25 − 7 = 18    What is 7 + 18?

# Topic 7 Place Value Workbook pages 21–23

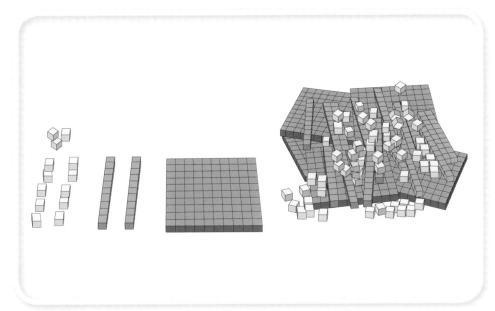

▲ What do the different colours show? How can they help you build numbers?

**Key Words**
place value
digits
tens
ones
hundred
expand
sum

In this topic, you will learn to use what you know about place value of tens and ones to help you add and subtract tens. You will also learn about place value of hundreds and use this to write, order and compare three-digit numbers.

## Getting Started

1 Look at the blocks in the photo. What number would be represented by:

a 7 yellows, 3 greens and a blue?    b 2 blues and a green?

c 2 blues and a yellow?              d 11 greens?

e 5 blues?

2 What blocks would you need to build these numbers?

a 90                                 b 190

c 109                                d 425

# Unit 1 Revising Tens and Ones

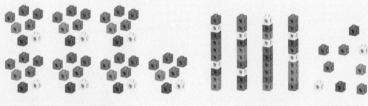

**Let's Think …**

James and Kennith have the same number of cubes.

- How many do they each have?
- Which ones did you count?
- Why?

▲ James' cubes     ▲ Kennith's cubes

*The position of each **digit** in a number tells us its **place value**.*

*39 = 3 tens and 9 ones*

*This is the same as 30 + 9.*

1 What is the number?

    a 4 tens and 5 ones     b 6 tens and 2 ones     c 8 tens and 9 ones

2 How many tens in each number?

    a sixty     b seventy     c twenty     d thirty

3 What is the value of the red digit in each number?

    a 38     b 93     c 88     d 79

4 Write the missing number in each number sentence.

    a 43 = ☐ + 3     b 29 = 20 + ☐     c 52 = ☐ + 2

**Looking Back**

a List all the two-digit numbers that can be made using these four digits.

      2    4    7    8

b Write your numbers in ascending order.

# Unit 2  Addition Using Place Value

**Let's Think …**

● What is ten more than:  30,  70,  88,  64?

You do not need to count on to add ones to numbers that end in 0.

In 10 + 2, the 2 means 2 ones.

The number 10 has 0 in the ones position, so you can replace 0 by 2.

10 + 2 = 12

You can use place value and number facts to add groups of tens quickly.

For example:     30 + 40 = 3 tens + 4 tens
                 3 + 4 = 7

So:              3 tens + 4 tens = 7 tens   or   70

1 Say the answers as quickly as you can.

a 10 + 3 =          b 10 + 5 =          c 10 + 6 =

d 1 + 10 =          e 7 + 10 =          f 8 + 10 =

2 Work these out without counting.

a 20 + 6 =          b 30 + 4 =          c 50 + 8 =

d 3 + 80 =          e 7 + 90 =          f 20 + 7 =

3 Add:

a 40 + 20 =          b 40 + 40 =          c 20 + 70 =

d 20 + 33 =          e 40 + 13 =          f 80 + 12 =

**Looking Back**

Add:

a 30 + 7 =      b 30 + 70 =      c 23 + 40 =      d 45 + 10 =

# Unit 3  Place Value to Hundreds

## Let's Think …

Each tower has 10 cubes.

- How many groups of 10 are there?
- Skip-count by 10s to find out how many cubes in all.
- What number is the same as 10 groups of 10?
- What number is the same as 12 groups of 10?

You already know that two-digit numbers have a tens place and a ones place.

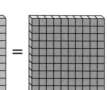

When you count above 99, you move from two-digit numbers to three-digit numbers.

The extra place is the **hundreds** place.
10 tens = 1 hundred.

We can show this on a place value chart like this:

| hundreds | tens | ones |
|----------|------|------|
| 1 | 0 | 0 |

In the number 100, there is 1 hundred, 0 tens and 0 ones.

These cubes show the number 139.

We can show this on the place value chart like this:

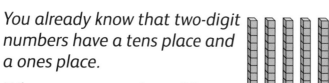

| hundreds | tens | ones |
|----------|------|------|
| 1 | 3 | 9 |

In the number 139, there is 1 hundred, 3 tens and 9 ones.

**1** Write the number shown by each set of cubes.

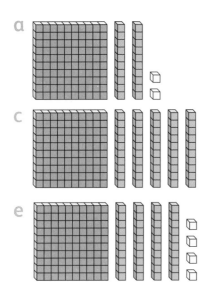

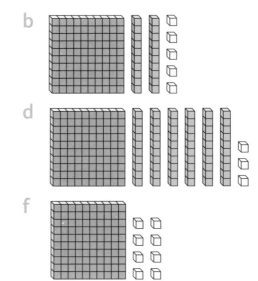

**2** What is the value of each underlined digit?

a **2**58          b 48**1**          c **6**73          d 5**0**9

**3** Which digit is in the hundreds place?

a 345          b 567          c 789          d 123

**4** Which digit is in the tens place?

a 543          b 765          c 709          d 230

**5** Write the set of numbers from Question 1 in ascending order. Now do the same with the sets of numbers in Questions 2, 3 and 4.

**6** True or false?

a 12 tens is less than 2 hundreds.

b 268 is more than 300.

c 303 is the same as 3 hundreds and 3 tens.

d 20 tens is more than 2 hundreds.

**7** How many three-digit numbers can you make using the digits 3, 4 and 5?

a Make a list of your numbers.

b Tick the numbers that are more than 400.

## Expanding Numbers

We can use place value to write any number as a sum.

For example: *264* = 200 + 60 + 4     *199* = 100 + 90 + 9

*This is called expanded notation. If you are told to* **expand** *a number, it means write it as a sum using place value.*

*When one of the digits is 0, it means there is no value in that place. We say the 0 is a place holder.*

*190* = 100 + 90 + 0     *There is 0 in the ones place, so the sum has no ones.*

*205* = 200 + 0 + 5     *There is 0 in the tens place, so the sum has no tens.*

1 Write the totals.

   a 300 + 10 + 4          b 200 + 50 + 8          c 200 + 90 + 7

   d 400 + 20 + 0          e 500 + 0 + 9           f 300 + 0 + 8

2 Expand each number. Write them in your copy book.

   a 369          b 521          c 209          d 862          e 420          f 999

3 Write these numbers using numerals.

   a eighty-five                    b six hundred thirty-five

   c four hundred three             d forty-seven

   e two hundred twelve             f nine hundred sixteen

   g fifteen                        h one hundred eleven

   i seven hundred seventy-seven    j six hundred forty

4 Jerome has made up a number game called *Building Numbers*. His rules are that you can only use the digits 1 and 0 and the **+** and **=** signs.

   This is how Jerome built 300.    100 + 100 + 100

   Work with a partner. Build these numbers using Jerome's rules.

   500, 275, 405, 266, 410

> We can use place value to order and compare numbers.
> 425 is greater than 325 because 4 hundreds are more than 3 hundreds.
> 425 is greater than 415 because 2 tens are more than 1 ten.
> 425 is greater than 421 because 5 ones are more than 1 one.

1 Order each set of numbers from smallest to greatest.

   a 131, 115, 126, 111, 119          b 316, 163, 631, 136, 613

2 Order each set of numbers from greatest to smallest.

   a 141, 161, 151, 171, 181          b 543, 345, 435, 534, 453, 354

3 Which number is greater in each pair? Write it in expanded notation.

   a 135 or 99          b 543 or 435          c 403 or 340

5 This machine is counting the cars driving over the Sydney Poitier Bridge. Two hundred sixty-three cars have already been counted.

   What is the next number that will be shown on the display with:

   a the same number of ones and tens as 263

   b the same number of ones and hundreds as 263

   c the same digit in the tens and hundreds place

   d the same digit in the hundreds, tens and ones place?

## Looking Back

Look at these number cards.

| 2 | 6 | 9 |

a What is the biggest three-digit number you can make with these three cards?

b What three-digit numbers can you make with these cards with 2 in the hundreds place?

# Topic Review

## What Did You Learn?

- The value of each digit in a number depends on its place or position.
- You can use place value to help you add numbers quickly and to compare and order numbers.
- A number with two digits has a tens place and a ones place.
- A number with three digits has three places: hundreds, tens and ones.
- 10 tens has the same value as 1 hundred.
- 345 can be expanded to 300 + 40 + 5.

## Talking Mathematics

Write down a three digit number. Give your group clues to help them guess your number. Talk about place value but do not name any of the digits in your number.

## Quick Check

1 Write these numbers in order from smallest to greatest.

345, 435, 305, 341, 353, 435

2 Write these in numerals.

a four hundred forty-three          b two hundred six

3 What is the value of the red digit in each number?

a 324          b 403          c 333          d 305

4 Expand these numbers.

a 125          b 99          c 390          d 407

5 What is 30 more than 125?

6 How many tens is 70 + 50? What is this in hundreds?

7 Write the next three numbers in each counting pattern.

a 220, 230, 240, ☐, ☐, ☐          b 700, 600, 500, ☐, ☐, ☐

c 109, 209, 309, ☐, ☐, ☐          d 935, 835, 735, ☐, ☐, ☐

8 How many days in a year? Write the number in expanded notation.

# Topic 8 Position Workbook pages 24–27

FINISH

## Key Words
position
ordinal numbers
grid
direction
left
right

▲ What is happening in this picture? Imagine you are the runner in blue and yellow. What would you say about this race?

Last year, you learned some position words and worked with ordinal numbers to give positions such as 1st, 2nd and 3rd. In this topic, you are going to learn to use words to write ordinal numbers. You will also revise position words and work with maps and grids to talk about position and find different objects.

## Getting Started

1 Describe the runner in the 2nd position.

2 Which two runners were closest together at the end of the race? In which positions did they finish?

3 Which runner finished in front of the runner in 5th position?

4 There were 10 runners in the race. In which position did the last runner finish?

# Unit 1  Ordinal Numbers

**Let's Think …**

If a person is number 5 in a row, we say they are in 5th position.

● Which word can we use for a person who is number 11 in a row?

● Which word can we use for a person who is number 17 in a row?

● Why?

We use **ordinal numbers** to describe the **position** of objects in an ordered group.

We can write ordinal numbers using numerals and words.

▲ first   second  third  fourth    fifth    sixth   seventh  eighth   ninth      tenth

The ordinal numbers after tenth, are:

| 11th | 12th | 13th | 14th | 15th |
|------|------|------|------|------|
| eleventh | twelfth | thirteenth | fourteenth | fifteenth |
| 16th | 17th | 18th | 19th | 20th |
| sixteenth | seventeenth | eighteenth | nineteenth | twentieth |

1 The map on page 53 shows the animals Tom saw as he walked home.

   a What was the last animal Tom saw before he got home?

   b Did Tom see the beetle first?

   c Which animal did Tom see fifth?

   d In which position was the ladybug?

e Where is the parrot?

f Which animal did Tom see third?

g Did Tom see the spider ninth?

h Which animal did Tom see seventh?

i Did Tom see the butterfly second?

j In which position was the bee?

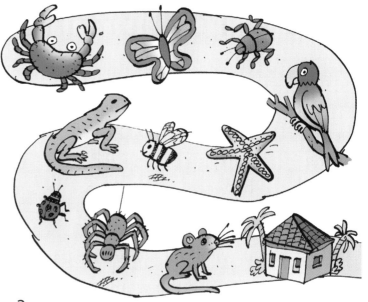

2 Write these positions as numerals in ascending order.

a thirteenth

b fourteenth

c eleventh

d twelfth

3 Tom marked the day he saw these animals on his calendar.

a Which month of the year is June?

b On which date did Tom see the animals?

c What day of the week was this?

d Which day is the sixth day of the week?

e What day of the week was 20 June?

| JUNE | | | | | | |
|---|---|---|---|---|---|---|
| S | M | T | W | T | F | S |
| | | | 1 | 2 | 3 | 4 |
| 5 | 6 | 7 | 8 | 9 | 10 | 11 |
| 12 | 13 | 14 | 15 | 16 | 17 | 18 |
| 19 | 20 | 21 | 22 | 23 | 24 | 25 |
| 26 | 27 | 28 | 29 | 30 | | |

**Looking Back**

Natasha finished 11th in a cross country race. Sharyn finished two places in front of her. Janae finished three places behind her.

a In which position did Sharyn finish?

b In which position did Janae finish?

# Unit 2  Position and Location

**Let's Think …**

Here are some of the words we use to describe positions.

| | | | | | |
|---|---|---|---|---|---|
| beside | next to | above | below | on | in |
| inside | outside | on top of | under | in front of | over |
| behind | between | near to | far from | up | down |

- Read the words with a partner.

- Talk about what each word means.

- Take turns to use the words to describe the position of objects in the picture. Try to use all the words.

## Left and Right

The words **left** and **right** are very useful for describing where things are.

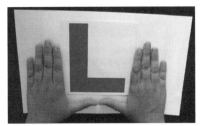

▲ Left hand    ▲ Right hand

You can use your hands to work out left and right. The thumb and pointer finger on your left hand make an L for left.

Move in the **direction** of your left hand to turn left.

Move in the direction of your right hand to turn right.

Answer these questions about Mrs Rolle's desk.

1 What is:

   a to the right of the diary

   b to the left of the ruler

   c in front of the tissues

   d behind the calculator?

2 Is the diary to the left or right of the class list?

3 Are the coins to the left or right of the class list?

4 Is the pen holder to the left or right of the jug?

5 Describe the position of the apple.

## Directions on Maps

*The position of places can be shown on a **grid**.*

*The grid below shows some places in Jason's neighbourhood.*

**Up**

| | | | | | | | |
|---|---|---|---|---|---|---|---|
| | | | | | | Park | |
| | Jason's home | | | | | | |
| | | | | | | Keith's home | |
| | | | | Straw market | | | |
| | Shops | | | | | Post office | |
| | | | | | | Library | |
| | | Butchery | | | | | |
| | | | | Sports club | | | |
| School | | | | | | Church | |

**Left**  **Right**

**Down**

**1** Follow these directions. Start at Jason's home each time. Where do you end up?

**a** Go 3 blocks right and 2 blocks down.

**b** Go 1 block down and 6 blocks right.

**c** Go 1 block left and 7 blocks down.

56

**2** Using the grid, write directions from the school to:

   **a** the sports club      **b** the library      **c** Keith's home.

**3** Jason's mom is at home. She goes to the shops, the butchery and the post office and then back home. Give one set of directions that could describe her journey.

The school bus has to travel from the bus stop to each of the three schools shown on this map.

**4** Suggest three routes the bus could take.

**5** Which route is the shortest?

**6** Write instructions that a new bus driver could follow to stop at all three schools.

**7** Swap routes with a partner. Follow your partner's route. Did you find your way to all the schools?

**Looking Back**

**1 a** What colour flower is third from the left?
  **b** What colour flower is fifth from the right?
  **c** What colour flower is between the pink flower and the blue flower?
**2** Write down directions for your route from your desk to the classroom door.

# Topic Review

**What Did You Learn?**

- Ordinal numbers like *first*, *second*, *10th* and *17th*, tell us the position of an object.
- We use words like *up*, *down*, *left* and *right* to describe the location of objects.
- A map shows the position of different objects.
- We use directions to find places on a map.

**Talking Mathematics**

Think about how you would tell a visitor to get from the school entrance gate to the principal's office. Remember to use the position and direction words you have learned in this topic. Share your directions with your group.

**Quick Check**

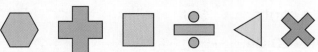

1 Look at the shapes. From the left, what colour is the:
   a 1st shape       b fifth shape       c eighth shape       d 7th shape?

2 What shapes are between the two blue shapes?

3 Which shape is:
   a furthest from the rectangle
   b to the left of the orange shape
   c to the right of the yellow shape?

4 What day of the week is the 20th of next month? Use a calendar.

5 Write the number that is 10 more than:
   a 20          b 100          c 212          d 435

6 Copy and complete:
   a $9 + \boxed{\phantom{0}} = 20$   b $5 + \boxed{\phantom{0}} = 20$   c $20 - 13 = \boxed{\phantom{0}}$   d $20 - 8 = \boxed{\phantom{0}}$

# Topic 9 Money

Workbook pages 28–30

## Key Words

coin
bill
dollar
cent
quarters
value
change
decimal point

▲ We use money all the time. When do you use coins? When do you use bills? How do you know which coins or bills to use?

You already know that we use Bahamian and American (US) coins and bills in the Bahamas. In this topic, you will learn to recognize different coins and bills. You will count and compare amounts of money made from different coins and bills. You will also learn how to write amounts of money in numerals and make up and solve problems involving money, including working out change.

## Getting Started

1 What money do we use in the Bahamas? Try to name all the coins and bills.

2 How does skip-counting help you to count amounts of money quickly?

3 Where do you see amounts of money written in daily life? How do you know they are amounts of money and not just numbers?

# Unit 1  Coins and Bills

**Let's Think …**

- What Bahamian coins are these?
- Say the name of the US coin that has the same value as each Bahamian coin.
- How are the coins from the Bahamas and the US different?

*Besides the 1¢, 5¢ and 10¢* **coins** *shown above, there is also a 25¢ Bahamian coin.*

- *The US coin that is worth 25* **cents** *is called a* **quarter**.

  *Can you think why?*

**1** Charles has these coins in his pocket.

60

a  Name each coin. Say whether it is a Bahamian or US coin.

b  What is the total of the 1¢ coins?

c  Do the 5¢ coins total 50¢?

d  What is the total of the 10¢ coins?

e  How many 25¢ coins have the same value as 1 dollar?

f  How much money does Charles have altogether?

g  Charles wants to buy four ice-creams for 2 dollars. Does he have enough money?

h  Each ice-cream costs 50¢. How many different ways can Charles make 50¢ with his coins?

2  These items on the sign are being sold at a school fair.

What combinations of coins can be used to pay for each item?

For sale:
Apples    35¢
Muffins   50¢
Juice     40¢
Bananas  44¢

3  List the coins that we use in the Bahamas in order, from greatest to least value.

4  Count the value of the coins in each set as quickly as you can. Write down each total.

a

b

c

d

e

f

5  Make each amount using a different set of coins.

*Paper money is known as notes or **bills**.*
*These are the Bahamian bills:*

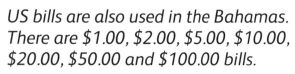 

*US bills are also used in the Bahamas.*
*There are $1.00, $2.00, $5.00, $10.00,*
*$20.00, $50.00 and $100.00 bills.*

*This is a US $1.00 bill:*

**6** Work out how much money each person has in total.

a I have twelve 5 dollar bills.

b I have nine 10 dollar bills.

c I have three 10 dollar bills and one 50 dollar bill.

d I have five $100.00 bills.

---

## Looking Back

**1** What bills could you use to make these amounts? Use as few bills as you can.

   **a** 50 dollars    **b** 8 dollars    **c** 225 dollars    **d** 80 dollars

**2** How do you know which country coins or notes come from?

**3** Maria finds a 2 dollar bill. How does she know it is not a Bahamian bill?

# Unit 2  Writing Amounts of Money

75¢  $75.00  $0.75  $7.50

We use the $ symbol to write amounts in **dollars**.

100 cents has the same value as 1 dollar. 100¢ = $1.00

You can write 115¢ using the dollar sign.

There are 100¢ ($1.00) and 15¢ cents left over. We write this as $1.15.

The **decimal point** separates the dollars from the cents.

5 dollars is written as $5.00 to show that there are no cents left over.

1  What is the highest number of cents you can have after the decimal point? Why?

2  Read these amounts.

    a $1.25   b $3.50    c $15.00    d $8.75    e $12.66    f $10.99

3  Write the amounts from Question 2 in order, from least to most money.

4  Are the amounts in words written correctly in numerals? Write *yes* or *no*.

    a one dollar thirty-three cents       $1.33

    b two dollars fifty-five cents       $25.5

    c one dollar       $100

    d forty-six cents       $0.46

    e twenty-five dollars       $0.25

**5** Write each total. Use a dollar sign and decimal point.

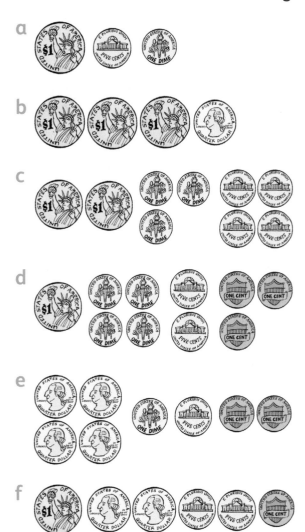

a

b

c

d

e

f

**6** Work out how much **change** you would get if you paid each amount in Question 5 with a 5 dollar bill.

## Looking Back

**1** List the coins and bills you could use to make each of these amounts.
$3.99, $4.58, $1.90, $2.09, $9.99

**2** Shenay pays one of these amounts with a 10 dollar bill. She gets $5.42 change. Which amount did she pay?

# Unit 3  Adding and Subtracting

## Let's Think …

Look at this section of the number chart.

| 1 | 2 | 3 | 4 | 5 | 6 | 7 | 8 | 9 | 10 |
|---|---|---|---|---|---|---|---|---|----|
| 11 | 12 | 13 | 14 | 15 | 16 | 17 | 18 | 19 | 20 |
| 21 | 22 | 23 | 24 | 25 | 26 | 27 | 28 | 29 | 30 |
| 31 | 32 | 33 | 34 | 35 | 36 | 37 | 38 | 39 | 40 |

- Find number 7.

  Which number is directly below it?

  How much more is that number?
- Find number 19.

  Which number is directly above it?

  How much less is that number?
- What pattern can you see?

We can add or subtract 10 without counting on or back.

For example:

23 + 10 = *33*        45 – 10 = *35*

$$\begin{array}{r} 23 \\ +\ 10 \\ \hline 33 \end{array} \qquad \begin{array}{r} 45 \\ -\ 10 \\ \hline 35 \end{array}$$

one ten more    one ten less

1 Do these without writing down any working out.

  a 12 + 10        b 25 + 10        c 36 – 10        d 48 – 10

2 Calculate:

  a    15¢          b    43¢          c    61¢          d    79¢
    + 10¢             + 10¢            – 10¢              – 10¢

3 Zara has 45¢. She gives 10¢ to her brother. How much does she have left?

4 Josh has 54¢. His mom gives him another 10¢. How much does he have now?

We can use place value to add two-digit amounts.
We can expand the numbers first and then add them.
For example:

| 34 + 25 | 34 is the same as | 30 + 4 |
| | 25 is the same as | 20 + 5 |
| | This gives | 50 + 9 |

*Look at this example of how to add together two-digit numbers mentally.*

14 + 23

Add the ones    4 + 3 = 7

Add the tens    10 + 20 = 30

Add the totals    30 + 7 = 37

Writing the addition in columns makes it easier to work out the totals.

```
  T  O
  4  1
+ 2  2
-------
  6  3     The answer is 63
```

You can use the same method to subtract.

```
  T  O
  4  7
- 2  3
-------
  2  4     The answer is 24
```

**1** Work these out. Remember to start with the ones column.

a
```
  T  O
  3  5
+ 2  2
```

b
```
  T  O
  5  4
+ 3  5
```

c
```
  T  O
  6  1
+ 1  5
```

d
```
  T  O
  5  8
- 2  6
```

e
```
  T  O
  7  6
- 3  1
```

f
```
  T  O
  9  4
- 1  2
```

**2** Write these calculations in columns then work out the answers.

a 50 + 32        b 18 + 61        c 39 + 30

d 87 – 23        e 99 – 47        f 88 – 50

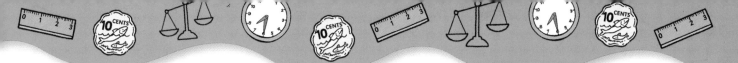

3 Michaela and Sharyn sell beads at a church market.

The poster shows their prices.

What is the cost of:

a 2 small beads

b 2 medium beads

c 2 big beads

d 2 bead packs

e 1 bead pack and
   1 string

f 1 big bead and
   1 medium bead?

### Price list

Big beads ......... 25¢
Medium beads .. 22¢
Small beads ..... 9¢
Bead pack ........ 60¢
Strings ............ 16¢

4 List the coins you would use to pay each
  amount in Question 3 exactly. Use as few
  coins as you can.

5 What change would you get from $1.00 if you bought:

a 1 bead pack

b 3 big beads

c 1 string and 2 medium beads?

6 Can you buy 2 big beads and 3 medium beads with $1.00?

7 You have $2.30 to spend on beads at the market. What would you buy?

## Looking Back

1 What is 10¢ more than:
   a 43¢          b 89¢          c 53¢          d 90¢?
2 What is 10¢ less than:
   a 43¢          b 89¢          c 53¢          d 90¢?
3 Nick spends 73¢ on bus fare and buys a sweet that costs 26¢.
   a How much does he spend altogether?
   b He started with $1.00. How much does he have left?

# Topic Review

**What Did You Learn?**

- Coins and bills come in different values.
- We use coins worth 1¢, 5¢, 10¢, 25¢ and 50¢ from the Bahamas and the US.
- There are Bahamian bills worth $1, $5, $10, $20, $50 and $100. We also use US bills.
- Amounts in cents are written with the cent symbol, for example, 45¢.
- Dollar and cent amounts are written with a dollar sign and a decimal point, for example, 1 dollar and 25 cents is written as $1.25.
- We can use place value to add two-digit numbers. We add the ones and the tens separately and then find the total.

**Talking Mathematics**

Which words mean the same as these phrases?

- A round piece of money worth half a dollar.
- Paper money with the number 10 printed on it.
- What you get back when an item costs less than the bill or coin you use to pay for it.
- A sign that looks like an S with lines through it.
- How much something costs.

**Quick Check**

1 Write 190 cents using the dollar sign.

2 List the coins you could use to pay 88¢. Use as few coins as possible.

3 Mike has a $5.00 bill and three $1.00 bills. He wants to buy a bicycle that costs $80.00. Explain why he cannot do that.

4 What is 62¢ plus 34¢?

5 Skip-count to find the total amount if you have 23 10¢ coins.

6 Sandra saves $5.00 every day for a week. How much does she have at the end of the week?

7 Write down three things that cost less than a dollar.

8 How many nickels have the same value as a quarter?

# Topic 10  Flat shapes Workbook pages 31–33

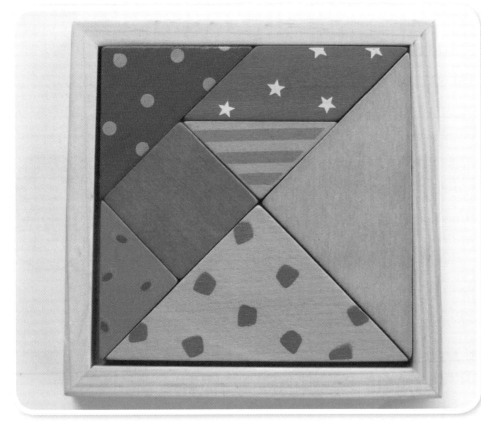

▲ How many different shapes can you find in this puzzle?
Can you name the shapes?

Workbook pages 31–33

<table>
</table>

**Key Words**

shape
flat
plane
circle
triangle
rectangle
square
side
corner
sort
classify

You should remember circles, triangles, squares and rectangles from last year. In this topic, you will learn more about the **properties** of different shapes. You will learn to use these properties to sort and classify shapes.

## Getting Started

1 Choose one shape from the photograph. Describe it without using its name.

2 Find a square and a rectangle in the photograph. What is the difference between a square and a rectangle?

3 What smaller shapes can you make by folding a square into two equal parts.

# Unit 1 Properties of Shapes

## Let's Think ...

Look around the classroom.

● Draw three different shapes that you can see.

● Which shape can you see most of?

---

**Flat shapes** *are also called* **plane** *or two-dimensional shapes.*

*The* **properties** *of a shape give us information about that shape, for example, about the number of* **sides** *or the number of* **corners** *that it has.*

*Sides are* <u>straight</u>.

side / This is a side.

side / Two sides meet at a corner.

▲ A **triangle** has 3 sides and 3 corners.

▲ A **circle** has no sides and no corners.

---

1 Here are some flat shapes.

a     b       c    d

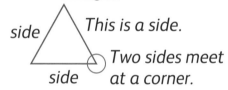

riantgel

angctreel

lericc

quaser

a Unscramble the names of the shapes and write them in a list.

b Next to each name, write the number of sides and corners that shape has.

2 Describe these flat shapes.

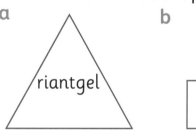

c

a

b

d

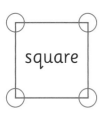

square

rectangle

▲ A **square** has 4 sides and 4 corners.

▲ A **rectangle** also has 4 sides and 4 corners.

*A shape with 4 sides is a square if the 4 corners are all the same size **and** the 4 sides are all equal in length.*

3 Clancy drew these shapes on squared paper.

a Which shapes are squares?

b Which shapes are rectangles?

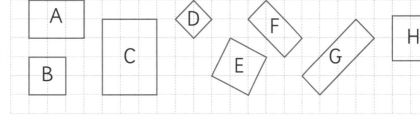

A    B    C    D    E    F    G    H

c What can you say about the sides and corners of a rectangle?

4 Draw a repeating pattern using one rectangle and three triangles as pattern units.

## Looking Back

1 Name these shapes.

a A shape with no corners and no sides.

b Two shapes with 4 sides.

c A shape with 3 sides and 3 corners.

2 Are there shapes with 4 sides and 4 corners that are not squares or rectangles?

# Unit 2  Sort and Classify Shapes

**Let's Think …**

Jayden sorted these shapes.

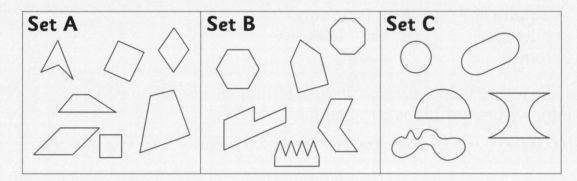

| Set A | Set B | Set C |

- What properties did he use to sort them?
- Can you think of any other ways to sort the shapes?

*When you **sort** shapes, you use their properties to put them into groups.*

*For example, you could use colour to sort shapes into red shapes and blue shapes.*

*When you **classify** a shape, you use its shape properties to work out what type of shape it is.*

*For example, you classify a shape as a triangle if it has 3 sides and 3 corners.*

1 Look at this Venn diagram carefully.

  a What shapes are in the left-hand circle?

  b What shapes are in the right-hand circle?

  c What shapes are in the overlapping part?

  d Why are some of the shapes not inside a circle?

  e Where would you put these shapes? Why?

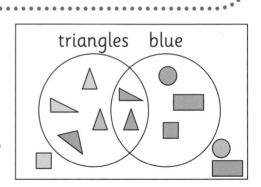

triangles    blue

A    B    C    D    E

**2** Work in groups. Record your answers in your copy books.

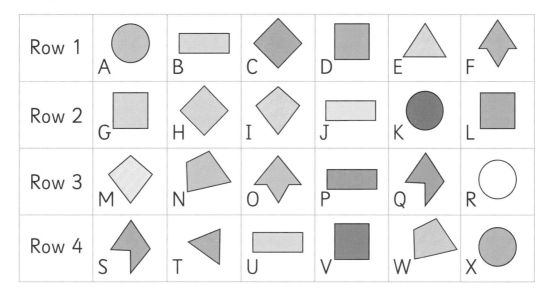

a Which shape is the 'odd one out' in the first row?

b Which shapes in the second row have 4 sides and 4 corners?

c How could you sort the shapes in the fourth row into three groups?

d Which shapes are triangles?

e Which shapes are squares?

f Which shapes have 4 sides but are not squares or rectangles?

**3** Which shapes match these descriptions?

a A shape with 6 sides that are not equal.

b A shape in the third row that has 5 sides.

c A flat round shape.

**4** Is shape O a triangle? Explain your answer.

**Looking Back**
Complete the sentences.
a This is not a rectangle because ...   b This is not a square because ...

# Topic Review

## Talking Mathematics

Work with a partner. Take turns to describe the properties of these shapes.

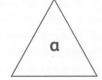

a    b    c    d    e    f

## Quick Check

1 Write the number that is made of three hundreds, eight tens and five ones.

2 Write the number that comes:

   a  before 529                b  after 529.

3 Copy and complete these number sentences.

   a  17 + 2 = ☐          b  23 + 10 = ☐          c  40 + 50 = ☐

4 Show how you can use place value to add 42 and 17.

5 How many corners does each shape have?

   a  a square

   b  a triangle

   c  a circle

6 How many of each shape are there in this box?

   a  triangles    b  circles    c  squares

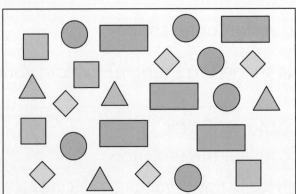

# Topic 11 Adding and Subtracting

Workbook pages 34–36

▲ Micah added 13 cents to 29 cents. The total was 42 cents. Explain how he could exchange the first two amounts for the second amount.

We can think about numbers as groups of tens and ones. We can also exchange ones for tens when we have more than 10 ones. In this topic, you are going to add and subtract numbers using what you already know about **place value** and writing numbers in expanded form.

## Getting Started

1 How can you combine these sets of cubes to get the total amount?

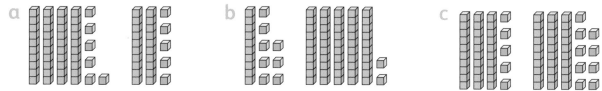

a      b      c

2 a What numbers do these cubes show?

 b How could you regroup the cubes to subtract the second number from the first number?

# Unit 1  Adding Two-digit Numbers

## Let's Think …

Selina modelled an addition sum using cubes.

- How could she write her sum?

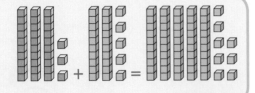

*To add two-digit numbers:*

- **Expand** *the numbers using place values.*
- **Add** *the tens and the ones separately.*
- *Combine them to get the* **total**.

*For example:*

| 32 | *is the same as* | 30 + 2 |
|---|---|---|
| + 26 | *is the same as* | 20 + 6 |
| 58 | | 50 + 8 |

1  Add these. Write each number in expanded form first.

   a  23 + 24       b  32 + 26       c  19 + 30

   d  36 + 42       e  16 + 63       f  45 + 54

   g  81 + 14       h  73 + 24       i  56 + 32

2  Add these. Show your working.

| a | | | b | | | c | | | d | | |
|---|---|---|---|---|---|---|---|---|---|---|---|
| | 3 | 2 | | 3 | 3 | | 7 | 0 | | 5 | 3 |
| + | 6 | 4 | + | 4 | 4 | + | 2 | 3 | + | 4 | 0 |

| e | | | f | | | g | | | h | | |
|---|---|---|---|---|---|---|---|---|---|---|---|
| | 1 | 7 | | 1 | 9 | | 2 | 7 | | 2 | 9 |
| + | 6 | 2 | + | 8 | 0 | + | 3 | 1 | + | 8 | 1 |

3  Ryan had 42¢. His mom gives him a quarter. How much does he have now?

## Regrouping

When you have more than 10 ones, you can **regroup** them into tens and ones.

Look at this example.

| 35 | 30 + 5 |
|----|--------|
| + 26 | 20 + 6 |
| 61 | 50 + 11 |
|  | 50 + 10 + 1 |

11 is the same as 10 + 1

= 61

4 Write each number in expanded form first and then add. Regroup then find the total.

a 46 + 35    b 39 + 22    c 37 + 8

d 38 + 16    e 49 + 26    f 35 + 45

Solve these problems. Show your working.

5 There are 28 students in Class A and 29 students in Class B. How many students is this altogether?

6 On Monday, 48 tourists visited a craft stall. On Tuesday, 37 tourists visited the stall. How many tourists is this in all?

7 Nadia has 35 stickers. Her friend has 16 more than her. How many do they have altogether?

**Looking Back**

Add. Show your working.

| a | 2 5 | b | 2 7 | c | 1 9 | d | 8 0 |
|---|-----|---|-----|---|-----|---|-----|
|   | + 4 0 |  | + 3 5 |  | + 7 2 |  | + 1 7 |

# Unit 2  Subtracting Two-digit Numbers

**Let's Think …**

- Sandy has 76 cubes. If she uses 42 of them, how many will she have left?
- How could you show this in writing?

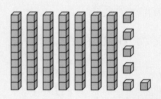

*To subtract two-digit numbers:*

- **Expand** *the numbers using place values.*
- **Subtract** *the tens and the ones separately. Subtract the ones first, then the tens.*
- **Combine** *them to get the answer.*

*Look at this example.*

*46 – 21*

*46 is the same as      40 + 6*

*21 is the same as    – 20 + 1*

                       ———————

                       *20 + 5 = 25*

**1** Subtract. Show your working.

| a T O | b T O | c T O | d T O |
|---|---|---|---|
| 3 8 | 6 4 | 8 6 | 9 7 |
| – 1 0 | – 2 0 | – 4 0 | – 6 0 |

**2** Do these subtractions in your copy book. Show your working.

| a T O | b T O | c T O | d T O |
|---|---|---|---|
| 5 8 | 3 4 | 8 7 | 6 5 |
| – 1 6 | – 2 1 | – 6 4 | – 5 2 |

| e T O | f T O | g T O | h T O |
|---|---|---|---|
| 4 9 | 7 6 | 9 5 | 4 9 |
| – 2 7 | – 4 3 | – 5 1 | – 3 6 |

## Exchanging

Sometimes you need to exchange 1 ten for 10 ones before you can subtract the ones.

Look at this example.
44 – 19

Think...
I do not have enough ones to subtract 9

Exchange 1 ten for 10 ones

$$\begin{array}{r} 40 + 4 \\ -\ 10 + 9 \\ \hline ? \end{array} \longrightarrow \begin{array}{r} 30 + 14 \\ -\ 10 +\ 9 \\ \hline 20 +\ 5 = 25 \end{array}$$

3 Exchange 1 ten for 10 ones.

a  40 + 4 = 30 + ☐

b  30 + 2 = 20 + ☐

c  50 + 5 = 40 + ☐

d  20 + 7 = 10 + ☐

e  40 + 0 = 30 + ☐

f  70 + 0 = 60 + ☐

4 Subtract. Show your working.

a  43 – 24

b  83 – 17

c  34 – 16

d  74 – 25

e  50 – 23

f  71 – 23

5 There are 60 students on a school trip; 37 of them are girls. How many are boys?

6 Sharon read a 64-page book. Michael read a 48-page book. How many more pages did Sharon read than Michael?

**Looking Back**
Subtract. Show your working.
a 50 – 30       b 55 – 41       c 51 – 45       d 60 – 34

# Topic Review

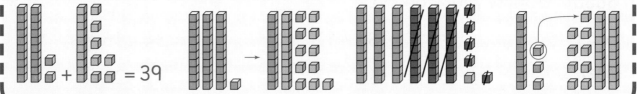

## What Did You Learn?

We can use place value to write numbers in expanded form.

- To add two-digit numbers:
  - Add tens and ones separately and combine to get the total.
  - Regroup to make a ten if you have more than 10 ones.
- To subtract two-digit numbers:
  - Subtract ones first, then tens.
  - Exchange 1 ten for 10 ones if there are not enough ones.

## Talking Mathematics

Match each statement with the correct picture.

| Regrouping | Adding tens and ones | Making a ten | Subtracting 35 from 66 |
| --- | --- | --- | --- |

$$+ = 39$$

## Quick Check

1 What is the missing number in each addition puzzle?

a   16
12   ○

b   17
8   ○

c   18
15   ○

d   19
11   ○

2 Add. Show your working.

a 12 + 34     b 50 + 43     c 42 + 58     d 19 + 23

3 Subtract. Show your working.

a 30 − 20     b 34 − 21     c 63 − 35     d 70 − 44

# Topic 12 Counting and Estimating

Workbook pages 37–39

▲ Guess how many marbles there are. Ten? 100? More than 100? How did you decide?

In this topic, you are going to learn how to make guesses in mathematics. You will find out how good guesses can help you make decisions, such as: is my answer sensible?

## Getting Started

1 How do you know, without counting, that there are more than ten marbles in the picture?

2 For each of the following, make a guess and write it down.

   a How many eyes are there in your classroom?

   b How tall is the classroom door?

   c How many passengers can sit in a bus?

   d How long is it until your bedtime?

3 Share your guesses with your group. Talk about how you guessed.

# Unit 1 Estimating

**Let's Think ...**

Janelle grabbed a handful of jelly beans.

- About how many beans is this?
- How did you decide this?

*An **estimate** is a **guess**. We estimate how much time something will take, how much money we will need, how tall things are and how many things there are in a group without having to measure or **count**.*

*You can use what you already know to help you make good estimates; for example, if you know your handspan is about 10 cm, you can compare items with your handspan and say that they are longer or shorter than this.*

1 Here are three groups of shells.

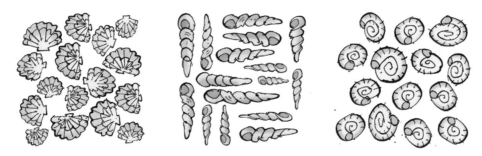

   a Which group has the most shells? Do not count them.

   b How many pink shells do you think there are?

   c Count the green shells. Do you think there are more or fewer yellow shells?

2 Which group do you think has the most faces? How could you check?

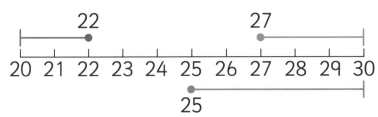

You can use an estimate to **check** whether your answer to a calculation is reasonable.

To estimate a calculation, first find the tens that are **closest to** the numbers you are working with.

*22 is closer to 20 than to 30.*     *27 is closer to 30 than to 20.*

*25 is halfway between 20 and 30. We use the higher ten, so we estimate 30.*

Estimate  23 + 39

  20 + 40 = 60       The answer will be close to 60.

Estimate  49 − 21

  50 − 20 = 30       The answer will be close to 30.

1 Estimate the answers to these calculations. Do not do the calculations.

  a 29 + 51        b 64 + 12        c 46 + 37        d 17 + 79

  e 82 − 49        f 64 − 19        g 77 − 53        h 38 − 24

2 Which is the correct estimation for 33 + 67?

  a 30 + 60            b 30 + 70            c 40 + 70

3 Which is the correct estimation for 86 − 44?

  a 100 − 40          b 80 − 50            c 90 − 40

**Looking Back**

1 a Estimate the number of students in your school.
  b Estimate how many are boys and how many are girls.
  c Explain how you estimated.
2 a Is 46 closer to 40 or 50?
  b Estimate 46 + 32.

# Topic Review

**What Did You Learn?**

- An estimate is a guess based on what you know.
- You can use an estimate to check whether the answer to a calculation is reasonable.

**Talking Mathematics**

- What is the difference between an estimate and a wild guess? Discuss this in groups.
- How can you improve your estimation skills for measuring?

**Quick Check**

1 Do you think these calculations are correct? Estimate, do not work them out.

| a | 19 | b | 89 | c | 90 | d | 28 |
|---|----|---|----|---|----|---|----|
| | + 37 | | − 52 | | − 27 | | + 72 |
| | 46 | | 41 | | 63 | | 90 |

2 The ladder in picture **a** is 2 m long.
Estimate the height of each tree.

a      b      c      d      e

3 Estimate first, then calculate.

a 27 + 53        b 19 + 54        c 82 + 8        d 12 + 43

e 75 + 27        f 36 + 29        g 43 − 21        h 55 − 42

i 63 − 29        j 45 − 35        k 84 − 34        l 54 − 35

4 Kieran has four coins in his hand. The total of the coins is 41¢.

  a Could two of the coins be quarters?

  b Could two of the coins be nickels?

  c What coins do you think he has? Show how you decided.

# Topic 13  Problem Solving

Workbook
pages 40–41

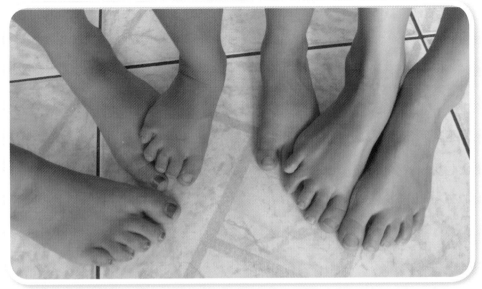

**Key Words**
read
list
decide
choose
strategy
pattern

▲ How many toes are there? Can you see all the toes? Does that change the answer? Why?

In this topic, you are going to learn some of the different ways you can solve problems involving numbers, patterns and money.

## Getting Started

Work in pairs. Talk about how you would solve each problem.

1 A cake costs 40¢ and a glass of juice costs 30¢. Anna spent exactly $1.70. What did she buy?

2 Mrs Gleeson has chickens and goats in her yard. There are 5 heads and 16 legs. How many chickens are there?

# Unit 1  Steps in Solving Problems

**Let's Think …**

Jessica and Maria have 14 sweets altogether.

● How many could they each have?

---

You can follow these steps to help you solve number problems.

*Step 1:* **Read** the problem carefully.

*Step 2:* **List** the information that is given. Highlight the important words and numbers.

*Step 3:* **Decide** what you have to do: add, subtract, share. Draw a line under what you have to find out.

*Step 4:* **Choose a strategy**. You might draw a picture, use a pattern, write a number sentence or do all three.

*Look at this example.*

*Gina has 18 fish in a pond. She adds 14 more. How many fish are in the tank now?*

$$18 + 14 = 10 + 8$$
$$+ 10 + 4$$
$$\overline{20 + 12} = 32$$

*Write the answer clearly.*     There are now 32 fish in the pond.

---

1 Drake has 26 crayons, but 9 are broken. How many of his crayons are not broken?

2 Sally bought a hairclip for 19¢ and a hairband for 31¢. How much money did she spend?

3 Daniel sees 29 cows in a field. 16 of the cows are brown. The rest are white. How many cows are white?

4 Make up a story problem of your own about this picture. Swap with a partner. Try to solve your partner's problem.

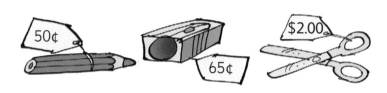

## Patterns in Problems

> Some problems can be solved using **patterns**.
>
> Ask yourself:
> - *What do I already know?*   ● *What do I need to find out?*
> - *Can I use a pattern to do this?*
>
> There are 40 toes under a table. How many people are sitting at the table?
>
> | **What do I know?** | **What do I need to find out?** | **Pattern** |
> |---|---|---|
> | 40 toes | How many people? | Count by 10s |
> | 10 toes per person | | 10 = 1 person   20 = 2 people |
> | | | 30 = 3 people   40 = 4 people |
>
> *If there are 40 toes, there are 4 people at the table.*

1 Kristen's mom is baking cupcakes for a party. Kristen wants an equal number of chocolate, vanilla and strawberry cupcakes. If 20 cupcakes are chocolate, how many cupcakes will there be altogether?

2 A shop-keeper kept a record of the toys she sold.

| Type of toy | Number sold |
|---|---|
| Computer games | 140 |
| Board games | 40 |
| Action/movie figures | 85 |
| Dolls | 25 |
| Bats and balls | 70 |

   a Which toy was most popular?

   b Which was least popular?

   c How many more action/movie figures were sold than dolls?

   d How many board games and bats and balls were sold altogether?

   e How many computer games and board games did she sell?

## Looking Back

a Make up two problems of your own using information from the table.

b Swap with a partner and solve each other's problems.

# Unit 2  Money Problems

**Let's Think …**

Anna and Jasmine want to buy one packet of sweets that costs $2.00.
Anna has 85¢ and Jasmine has 95¢.

● Do they have enough money?

● Anna's mom gives them a quarter. Do they have enough money now?

● Tell your group how you found your answers.

1 Mr and Mrs Bullard take their children Amy, aged 12, and Mark, aged 8, to the amusement park. They see this sign at the entrance.

| Tickets | | |
|---|---|---|
| | Age 10+ | Age 3–9 |
| Fun Park | $20.00 | $15.00 |
| Water World | $25.00 | $10.00 |
| Wild Creatures | $15.00 | $8.00 |

  a Mark wants to go to the Fun Park. How much would his ticket cost?

  b Amy wants to go to Wild Creatures with her mother. How much would their tickets cost?

  c Mr Bullard wants to go to all of the parks. How much would his tickets cost?

  d Mrs Bullard thinks the whole family should go to Water World. How much will these tickets cost?

  e Mr Bullard uses four Bahamian bills to pay the exact total for the tickets to Water World. What bills does he use?

  f Mrs Bullard pays for her and Amy to go to Wild Creatures using a $50.00 bill. How much change should she get?

  g Mr Bullard gives Mark $20.00. Which parks could he visit?

**Looking Back**

1 How much would it cost for your group to visit the Water World park?

2 Make up a story problem of your own using these prices. Show how to solve your problem.

# Topic Review

**What Did You Learn?**

● To solve a problem: read it carefully, list what you know, decide what you have to do, choose a strategy to solve it.

**Talking Mathematics**

You have been asked to teach a new student how to solve story problems. What would you tell them?

**Quick Check**

1 18 toffees cost 30¢. How many toffees could you buy with 15¢?

2 Sandy can see people and dogs on the beach. She counts 14 heads and 32 legs. How many people can she see on the beach? How many dogs can she see on the beach?

3 Jemma has 25¢. After she gets paid for doing some chores, she has $2.20. How much did she get paid for the chores?

4 Micah wants to buy a book that costs $3.50. He has two 1 dollar bills and eight quarters. Does he have enough money?

5 Josh has $3.50. He gets paid $2.00 for washing the car and his gran gives him a quarter, two dimes and a nickel from her purse. How much does he have now?

6 Nick is 2 years older than James. James is 4 years younger than Mark. Mark is 12. How old are Nick and James?

7 Match the amounts of time.

| 7 days | 365 days | 12 months | 24 hours |
| leap year | day | year | week |

8 a Describe this counting pattern.
   380, 370, 360, 350, 340

  b What are the next five numbers in the pattern?

  c What could this person be counting?

# Topic 14  Graphs Workbook pages 42–45

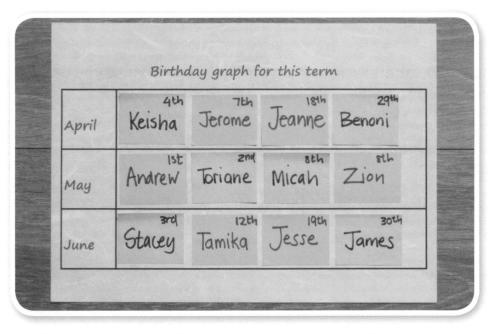

Birthday graph for this term

| | | | | |
|---|---|---|---|---|
| April | 4th Keisha | 7th Jerome | 18th Jeanne | 29th Benoni |
| May | 1st Andrew | 2nd Toriane | 8th Micah | 8th Zion |
| June | 3rd Stacey | 12th Tamika | 19th Jesse | 30th James |

**Key Words**

collect
data
concrete
pictograph
symbol
bar graph
horizontal
vertical
conclude
predict
range
mode

▲ Class 2 made this display. What does it show? How is it organized?

Do you remember what you learned last year about pictographs and block graphs? In this topic, you are going to answer questions about these graphs and collect information to draw your own graphs. You will also learn how to find the mode and range of sets of data.

## Getting Started

1 Look at the class display of birthdays in the photograph.

   a How could you show this information on a pictograph?

   b How could you show this information on a block graph?

2 In which month did most students have birthdays?

3 How many students had birthdays in:

   a May            b June?

4 Were there fewer birthdays in April or May?

# Unit 1  Revisiting Pictographs

## Let's Think ...

The students in Class 2 stuck stickers onto a chart to show their favourite fruits.

- Say four things that you can tell from this chart.
- Is this a graph? How do you know?

When you **collect** information (**data**) about a particular subject, you can organize it and display it on a graph.

There are different types of graphs.

The chart above is called a **concrete** graph. This type of graph uses real items to show the **information**. On this graph, the students used stickers. On the concrete graph showing birthdays, they used sticky notes.

A **pictograph** is similar to a concrete graph but it uses small pictures or **symbols** to show the information. A pictograph should have a key to tell you what each picture or symbol represents.

This pictograph shows the same information about favourite fruits as the concrete graph.

Favourite fruits

| Pineapple | 𝓍 𝓍 𝓍 |
| Mango | 𝓍 𝓍 𝓍 𝓍 𝓍 𝓍 |
| Banana | 𝓍 𝓍 𝓍 𝓍 𝓍 |
| Plum | 𝓍 𝓍 𝓍 𝓍 |

Key
𝓍 = 1 person

We can use graphs to find answers to questions (draw **conclusions**) and to **predict** what might happen in the future.

1 Look at this graph and answer the questions.

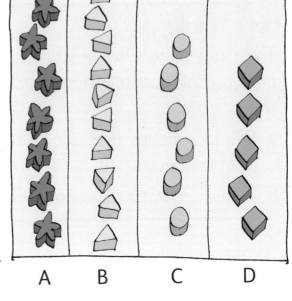

  a Which column has the most shapes?

  b Which column has the fewest shapes?

  c How many purple blocks are there?

  d Which column has twice as many shapes as the blocks column?

  e Which column has more than 6 but less than 8 shapes?

  f Which columns have an odd number of shapes?

  g What do you call this type of graph?

A    B    C    D

2 A group in Class 2 collected information from their friends and used it to draw this graph.

  a What type of graph is this?

  b What does this graph show you?

  c What conclusion might you make about boys and fruit?

  d Can you predict who will eat healthy food when they are older?

  e How many more boys than girls like juice?

  f What title could you give this graph?

**Looking Back**

What is the difference between a concrete graph and a pictograph?

# Unit 2  Bar Graphs

**Let's Think …**

- How many candies are there of each colour?
- Record the data in your copy book.

You can collect data by counting objects in different groups like you did with the candies.

You can record your information in a table like this one:

| Colour | Number | Number |
|--------|--------|--------|
| Red | IIII IIII | 10 |
| Blue | IIII IIII | 10 |
| Yellow | IIII III | 8 |
| Green | IIII IIII | 9 |
| Purple | IIII IIII I | 11 |

You can make small marks called tallies as you count each item. One mark (I) means 1 item. Every 5th mark is drawn across the first 4 to show a group of 5. Then we can skip-count by 5s.

Once you have collected and organized your data, you can draw a **bar graph**.

This bar graph shows the information in the table.

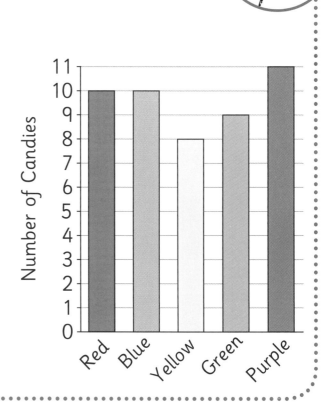

1 Draw a bar graph to show the colours of candies in the photograph. Use the data you wrote in your copy book.

**2** Mr Butler collected information about his students' favourite colours and drew a graph of the data.

a Which colours were chosen by the same numbers of students?

b How many students are there in the class?

c There are 15 girls in the class. How many boys are there?

d Five girls chose silver as their favourite colour. How many boys chose silver?

e An equal number of boys and girls chose red. What is the number?

f What type of graph is this?

g What is the title of the graph?

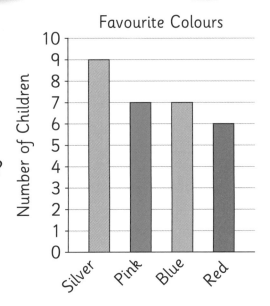

Favourite Colours

---

## Looking Back

Natalie's dad recorded how much television Natalie watched each week for 4 weeks in a table.

| Week | 1 | 2 | 3 | 4 |
|------|---|---|---|---|
| Hours of television | 7 | 10 | 9 | 8 |

**a** Which graph shows the data correctly?

Natalie's TV Viewing

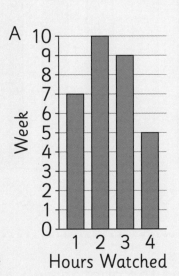

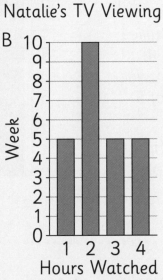

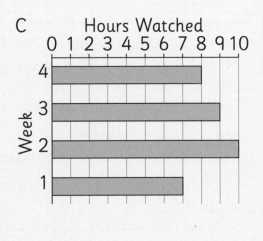

**b** Why are the other two graphs incorrect?

# Unit 3  The Range and the Mode

## Let's Think ...

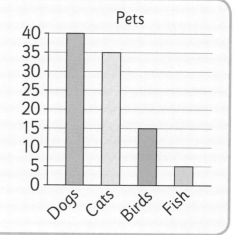

Pets

- What does the graph show?
- Which is the most popular pet?
- How many people have the most popular pet? How many people have the least popular pet? What is the difference between these numbers?

---

The item that occurs most often in a set of data is called the **mode**.

- You can find the mode on a bar graph by looking for the longest bar. In the graph above, the mode is dogs.

The **range** is the difference between the highest and lowest values in the data.

- You can find the mode on a bar graph by subtracting the value of the lowest bar from the value of the the highest bar. The range of pets is 40 − 5 = 35.

---

**1** Nat and Lily each rolled a dice. Find the mode of each person's scores.

Nat                                           Lily

**2** These are the heights of 10 boys and 10 girls.

| Boys | 120 | 140 | 130 | 120 | 130 | 110 | 120 | 150 | 140 | 140 |
|------|-----|-----|-----|-----|-----|-----|-----|-----|-----|-----|
| Girls | 110 | 100 | 110 | 130 | 130 | 140 | 110 | 120 | 130 | 120 |

a What is the mode for boys?   b What is the range for girls?

## Looking Back

These are the points scored in five matches: 40, 45, 45, 39, 48

a Which score is the mode?   b What is the range of the scores?

# Topic Review

## Talking Mathematics

Match the words to the correct meanings.

| | |
|---|---|
| concrete graph | the most popular item in a set |
| bar graph | the highest value minus the lowest value |
| range | a graph with a scale and bars of different lengths |
| mode | a graph made out of real objects |

## Quick Check

1 Class 2 recorded the colours of 40 cars passing the school.

| Colour | White | Red | Silver | Black | Other |
|---|---|---|---|---|---|
| Number of cars | 17 | 5 | 10 | 2 | 6 |

  a Draw a pictograph to show the information.

  b Which colour is the mode?

2 The tallest girl in a class is 125 cm. The shortest girl is 86 cm. What is the range of heights?

3 Look at the graph carefully.

  a Which was the most popular colour of T-shirt?

  b How many students wore purple T-shirts?

  c How many fewer students wore red T-shirts than orange T-shirts?

  d Is orange the mode of the data? Why?

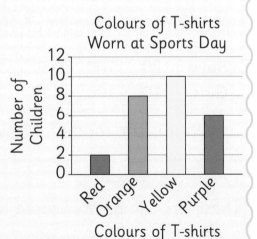

Colours of T-shirts Worn at Sports Day

# Topic 15 Fractions Workbook pages 46–48

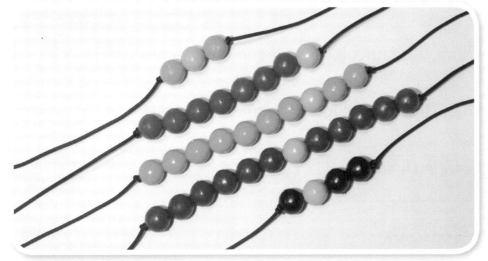

Workbook pages 46–48

**Key Words**
fraction
part
whole
equal
half
third
fourth
eighth
tenth
twelfth

▲ How many beads on each string? Look at the top string. One out of the three beads is yellow. Describe the yellow beads on the other strings in the same way.

**Equal parts** of a **whole** item and **equal shares** of a group of items are called fractions. Last year, you worked with halves, thirds and fourths. In this topic, you will practice working with these fractions and learn to recognize, draw and name some other fractions. You will also work with questions and problems that contain fractions.

## Getting Started

1 Which of these foods have been cut into halves? How do you know the other foods are not cut in half?

2 How many parts have these shapes been divided into? What would you call each part? Why?

   a

   b

   c

   d

97

# Unit 1 Revisiting Fractions

**Let's Think …**

- How would you share this chocolate bar equally between two people?
- How would you share this chocolate bar equally between three people?
- What would you call each share?

*Do you remember these fraction names?*

$\frac{1}{2}$

$\frac{1}{3}$

$\frac{1}{4}$

one **half**

one **third**

one **fourth**

$\frac{1}{4}$ *means 1 out of 4* **equal parts**.

**1** What fraction of each shape is shaded?

a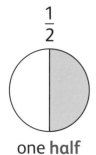

b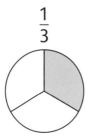

c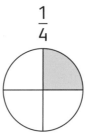

d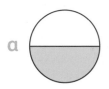

e

f

g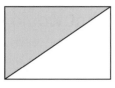

h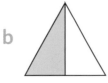

2 What fraction of each group is circled?

a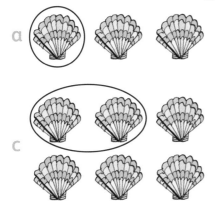

b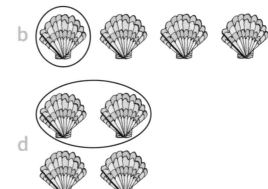

c

d

3 Ms Newton has a box of juicy fruits.

   a If she shares the juicy fruits equally with one friend, how many will each person get?

   b How could she share the fruits equally between three people?

   c What is the greatest number of people that could share the juicy fruits and get an equal share?

4 Look at this group of fish.

   a How many fish are in half the group?

   b What fraction of the group is 4 fish?

   c A fisherman catches 3 of the fish. What fraction of the group is this?

5 Heston has two 50¢ coins.

   a How much money does he have in all?

   b He gives half the money to a charity. How much is this?

   c How much is one fourth of his money?

---

**Looking Back**

Fold a sheet of paper to show:

**a** halves          **b** fourths          **c** thirds.

Can you find more than one way to fold the sheet each time?

# Unit 2  More Fractions

**Let's Think ...**

- $\frac{1}{2}$ means 1 out of 2 equal parts.

- $\frac{1}{4}$ means 1 out of 4 equal parts.

- What do you think $\frac{1}{10}$ means? Why?

Look at these fractions.

| $\frac{1}{8}$ | $\frac{1}{8}$ | $\frac{1}{8}$ | $\frac{1}{8}$ | $\frac{1}{8}$ | $\frac{1}{8}$ | $\frac{1}{8}$ | $\frac{1}{8}$ | eighths |

In the top row, each part is 1 out of 8 equal parts of the rectangle.

The bottom number in a fraction tells us how many equal parts the **whole** is divided into.

1 Complete these sentences.

    a One third is one out of ___ equal parts.

    b One fourth is one out of four ___ parts.

    c One ___ is one out of eight equal parts.

    d One tenth is ___ out of ten equal parts.

    e One ___ is one out of twelve equal parts.

2 Look at the cake on the next page. Megan could not decide which flavours she wanted on her birthday cake, so her dad got a cake with equal amounts of vanilla, strawberry, chocolate and lime.

    a There are four flavours. How has the cake been divided to show the four flavours?

b How many slices are there altogether?

c How would you write the fraction for one slice of the cake?

d Megan ate a slice of chocolate cake. How many parts of the cake are left?

e Megan's mom then ate a slice of lemon cake. What fraction of the cake is left now?

3 For the school cake sale, Megan's mom baked a cake in the shape of a rectangle and cut it into equal slices.

a How many equal parts are there?

b What fraction of the cake is one slice?

c Five slices of cake are sold. What fraction is left?

d One slice of cake costs 20¢. How much would three slices cost?

e What fraction of the cake could you buy with $1.00?

f Draw your own rectangular cake. Show how you could divide it into twelfths. Write the fraction on one of the slices.

---

**Looking Back**

What fraction of each shape is each colour?

a        b                                    c

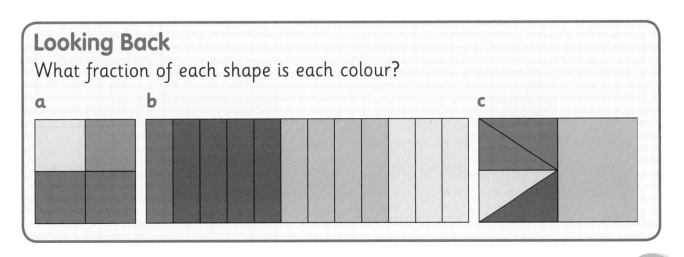

# Topic Review

## What Did You Learn?

- A fraction is an equal part of a whole or group.
- $\frac{1}{2}$ means 1 out of 2 equal parts.
- In a fraction, the bottom number tells us how many parts the whole is divided into. The top number tells us how many parts we have.

## Talking Mathematics

Use what you have learned about fractions to explain what each person means.

> I give a tenth of my wages to charity.

> Half the people on the beach got sunburnt.

> Cut the pizza into eight slices.

## Quick Check

1 There are eight objects in a set. How many is:

a $\frac{1}{2}$ of eight    b $\frac{1}{4}$ of eight    c $\frac{1}{8}$ of eight?

2 What fraction of each shape is shaded?

a 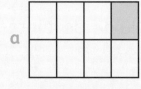    b    c

d     e 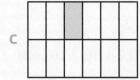    f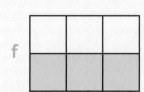

3 How many 10¢ stamps could you buy with 50¢?

4 a Write seven hundred twenty-five in numerals.

   b Write the number that is 100 less than seven hundred twenty-five.

   c Write the number that is 5 more than seven hundred twenty-five.

5 Joe says half the year has already past. What month is it now?

# Topic 16 Measuring Length and Height Workbook pages 49–51

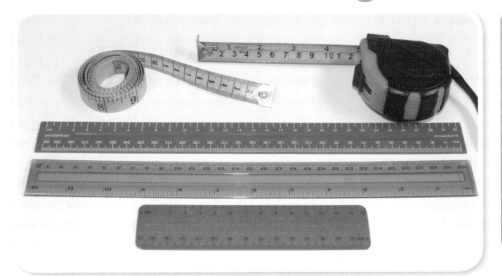

▲ What could you measure with each of these different measuring tools? Which units would you measure in?

You have already worked with centimetres and metres and, in this topic, you will also learn about decimetres. A lot of the work you do in this topic will be practical. You will practice measuring in standard units and learn to use what you already know to estimate before you measure.

## Getting Started

1 Try to picture a centimetre in your mind.

   a Is your thumbnail a centimetre long?

   b Is your thumb longer or shorter than a centimetre?

   c Name some items that are about a centimetre long.

2 Try to picture a metre in your mind.

   a Is your arm a metre long?

   b Is your table a metre long?

   c Are you shorter or taller than a metre?

# Unit 1 Centimetres and Metres

## Let's Think …

● What units would you use to measure:
  ● the length of your pencil
  ● the length of the classroom
  ● the height of your desk
  ● the height of a door?
● Why?

We use different **units** and measuring instruments to **measure** different **lengths** and **heights**.

We can use a ruler to measure short lengths in **centimetres** (cm).

We can use a tape measure or metre stick to measure longer lengths in **metres** (m).

```
0  10  20  30  40  50  60  70  80  90  100
1 METRE
```

*100 centimetres = 1 metre*

When you write down a measurement, you must also write down the units that you used to measure, for example, 10 cm or 3 m.

**1** Are these measurements reasonable? Why?

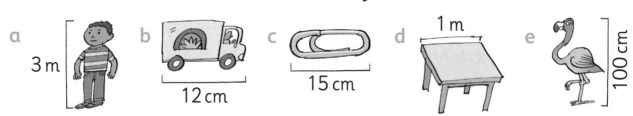

a 3 m   b 12 cm   c 15 cm   d 1 m   e 100 cm

## Looking Back

**1** Write down three items that are about 1 m long.

**2** Measure three items in the classroom in centimetres.

# Unit 2 Estimating

*You can use an estimate to decide whether or not you have measured correctly.*

*To **estimate**, think about lengths that you already know and compare them with what you are measuring.*

*I know a ruler is 30 cm long.*

*I think this table will be about 2 rulers high.*

*I estimate the height is 60 cm.*

*I have a metre stick.*

*I think 4 metre sticks will stack against the wall.*

*I estimate the height of the room is 4 m.*

1 The red line above the ruler is 5 cm long. Use this to help you estimate the length of the other lines.

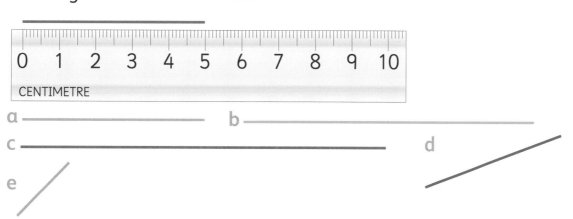

**2** Estimate the length of each animal in centimetres. Write down your estimate and then measure each animal. Write your measurements next to your estimates.

**3** Choose the best estimate for each measurement.

   About 1 m          About 5 m          About 10 m

   a  the width of a door
   b  your height
   c  the length of the whiteboard
   d  the length of a car
   e  the width of the passage
   f  the length of your table
   g  the height of your school building
   h  the width of your classroom
   i  the height of a mango tree
   j  a distance of ten steps

**4** Choose three measurements from the list. Use a metre stick or tape measure to measure each length to the nearest metre. Record your measurements.

**Looking Back**
Write down the objects in your classroom that fit each estimate.
   **a** 10 cm          **b** 1 m          **c** 2 cm          **d** 2 m

# Unit 3  Decimetres

## Let's Think …

There are 100 cm in 1 m.

- How many centimetres in half a metre?
- What fraction of a metre is 25 cm?
- How many pieces 10 cm long can you cut from a 1 m piece of string?

*A metre can be divided into equal parts.*
*This metre stick has been divided into ten equal parts.*
*Each part is $\frac{1}{10}$ of a metre.*

*A unit of 10 cm is called*
*a **decimetre**.*

*10 cm = 1 dm*

*10 dm = 1 m*

| $\frac{1}{10}$ | $\frac{1}{10}$ | $\frac{1}{10}$ | $\frac{1}{10}$ | $\frac{1}{10}$ | $\frac{1}{10}$ | $\frac{1}{10}$ | $\frac{1}{10}$ | $\frac{1}{10}$ | $\frac{1}{10}$ |

1 METRE

1 Make your own decimetre measure using a strip of card or a piece of string divided into centimetres.

2 Use your decimetre strip to measure the length of each item to the nearest decimetre. Remember to estimate before you measure.

- a the length of a pen
- b the height of a chair
- c the length of your foot
- d the width of this book
- e the length of a plastic ruler
- f the height of a trash can

## Looking Back

Find an item in your classroom that is:

**a** shorter than a decimetre    **b** about a decimetre long

**c** about 10 dm long.

# Topic Review

**Talking Mathematics**

These instructions for measuring the length of a desk have been jumbled up.

What is the correct order?

A Record your measurement.

B Use a tape measure marked in centimetres and metres.

C Remember to write the units.

D Estimate the length before you measure.

E Start measuring from the 0 mark on the tape measure.

**Quick Check**

1 Estimate and then measure the length of each bar. Which bar is about 1 dm long?

a

b

c

2 List four lengths you would measure in metres.

# Topic 17 Counting and Comparing Numbers

Workbook pages 52–54

## Key Words
count
before
after
pattern
compare
order
less than
greater than
symbol

▲ What number is shown on the screen? What is the value of each 9 in this number?

You can already count well up to 999. In this topic, you are going to learn some new symbols that you can use when you compare numbers. You are also going to find missing numbers in number charts and skip-counting counting patterns.

## Getting Started

1 Skip-count in 2s from 412 to 450.

2 Which even number comes after 466?

3 Which odd number comes before 478?

4 Which is smaller: 456 or 465?

5 Which is bigger: 498 or 500?

6 What is this counting pattern? 405, 407, 409, 411, 413, 415, 417

# Unit 1 Comparing Numbers

**Let's Think …**

● Which is greater: 665 or 655?
● Which is less: 689 or 698?

We need to **compare** numbers to say which is greater or smaller.

We can use **symbols** instead of words to write **greater than** and **less than** in a shorter way.

*<* means less than.

For example:          4 < 5 and 356 < 365

*>* means greater than.

For example:          5 > 4 and 365 > 356

You can remember the symbols by thinking about an alligator's mouth.

The open mouth always points towards the bigger number.

**1** Read each statement.

  a 12 > 5          b 123 < 125          c 188 > 88
  d 13 < 16         e 456 > 446          f 900 < 999

**2** Jayden has compared some numbers. Check his work. Has he used the correct symbol?

  a 2 < 6               b 22 > 20             c  48 > 72
  d 99 < 89            e 10 > 5 + 5         f  50 + 8 < 60
  g 23 + 3 < 20 + 6   h 51 > 15            i  81 > 40 + 40

**Looking Back**

Read each statement forwards and then backwards. What do you notice?

  **a** 17 < 19     **b** 200 < 300     **c** 876 > 867     **d** 900 < 999

# Unit 2 Count, Order, Compare

**Let's Think ...**
- Skip-count by 5s from 505 to 580.
- Count back from 800 to 780.
- Which of these are odd numbers? 662, 664, 666, 667

*You can **count** forwards and backwards and you can skip-count.*
*You can **order** numbers by size.*
*Ascending numbers go from smallest to greatest.*
*Descending numbers go from greatest to smallest.*
*You can use the symbols < or > when you compare numbers.*

1 Start at 608 and skip-count by 2s to 620. Did you count any odd numbers?

2 Skip-count back by 10s from 800 to 700. What is the same about all the numbers?

3 Write each set of numbers in ascending order.

   a 234, 432, 324, 243, 423, 342

   b 589, 895, 985, 859, 958, 598

4 Copy these number pairs. Write < or > in the box to make each statement true.

   a 324 ☐ 423     b 523 ☐ 532     c 802 ☐ 820

   d 189 ☐ 289     e 800 ☐ 720     f 999 ☐ 998

## Looking Back
What are the missing numbers?

| 100 less | Number | 100 more |
|---|---|---|
| | 800 | |
| | 206 | |
| | 344 | |
| | 612 | |

| 100 less | Number | 100 more |
|---|---|---|
| | 555 | |
| | 723 | |
| | 399 | |
| | 110 | |

# Topic Review

**What Did You Learn?**

- The symbol < means less than.
- The symbol > means greater than.

**Talking Mathematics**

Complete each sentence. Use one example from daily life.

- We put things in size order when we …
- We count when we …
- We compare amounts when we …

**Quick Check**

1 Write the missing numbers. What is the pattern in each set?

  a  90, 190, 290, ☐, ☐, ☐, ☐, ☐

  b  912, 812, 712, ☐, ☐, ☐, ☐, ☐

  c  987, 988, 989, ☐, ☐, ☐, ☐, ☐, ☐

  d  973, 972, 971, ☐, ☐, ☐, ☐, ☐, ☐

2 Lisa spent 36¢. She paid with a 50¢ coin and received 24¢ change. Did she get the correct change?

3 Sam had three coins totalling 75¢.

  a  What coins could he have had?

  b  He spends 49¢. How much does he have left?

4 Compare these numbers. Write three statements using < or >.
123, 321, 231, 312

5 What is:

  a  10 less than 456

  b  100 less than 800

  c  10 more than 345

  d  100 less than 350?

# Topic 18 Symmetry Workbook pages 55–57

**Key Words**
symmetry
line of symmetry
identical
half
mirror image

▲ Compare the two sides of the photograph. What do you notice?

In this topic, you are going to investigate symmetry. We say a shape has symmetry if you can draw a line that divides it into two **identical** parts, like the red line on the butterfly. If you could fold the photograph along the red line, the two parts of the butterfly would match exactly.

## Getting Started

1 Fold a piece of paper in half and draw a shape along the folded edge.

2 Cut along the lines of the shape. Do not cut along the fold.

3 Unfold the shape. What do you notice?

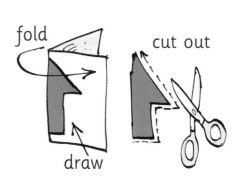

# Unit 1 Lines of Symmetry

## Let's Think ...

- Look at the starfish.
- Use a ruler to show where you could draw a line to divide it into two equal parts.

*If you can draw a line on a shape that divides it into two **identical** parts, we say the shape has **symmetry**. The line is called a **line of symmetry**.*

*The two parts of the shape match exactly. We say that they are the **mirror image** of each other. If you place a small mirror on the line of symmetry, you can see the whole of the original shape. **Half** the shape is on the paper, the other half is in the mirror.*

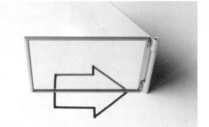

*If you can not see the original shape, then you have not placed the mirror on a line of symmetry.*

*Look carefully at this example.*

**1** Say whether each line is a line of symmetry or not.

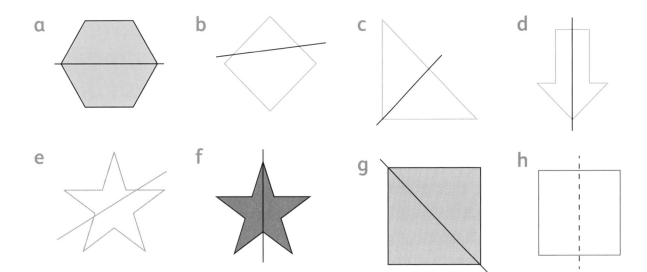

a    b    c    d

e    f    g    h

**2** Find your initials in this list. Decide whether or not the letters of your initials are symmetrical. You can use a ruler or a mirror to help you.

# A B C D E F G H I J K L M
# N O P Q R S T U V W X Y Z

Write your intials and draw the lines of symmetry on the symmetrical letters.

## Looking Back
The red lines on these pictures are **not** lines of symmetry. Explain why not.

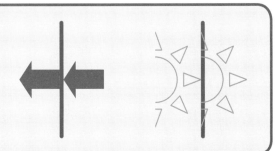

# Topic Review

**What Did You Learn?**

- A shape has symmetry if you can draw a line that divides it into two identical parts. The two parts are the mirror image of each other.
- The line that divides a shape into two identical parts is called a line of symmetry.

**Talking Mathematics**

These students are talking about symmetry, but they have not used the correct words. Find the wrong words and correct each statement.

> A line of mystery divides a shape into two parts.

> When a shape is symmetrical, the two thirds are identical.

> A mirror image shows the back side of a shape.

**Quick Check**

1 Which of these objects have symmetry?

2 Micah has $5.00. How much more does he need to make $20.00?

3 There are 20 cars parked at a school.

$\frac{1}{2}$ are white, $\frac{1}{4}$ are black, 1 car is red and the rest are silver.

How many cars are:

a white          b black          c silver?

# Topic 19 Telling the Time

Workbook pages 58–60

## Key Words

morning
afternoon
evening
night
a.m.
p.m.
noon
midday
midnight
o'clock
half-past
quarter-past

▲ What is the time on the clock? Can you tell whether this is in the morning or the evening?

In this topic, you are going to revise some words we use to talk about different times of the day and learn some new ones. You will also practice telling the time on the hour and half hour and learn to tell the time to the quarter hour.

## Getting Started

1 List three things that you do:

  a in the morning

  b in the afternoon

  c at night.

2 Kayleigh says that she gets up at 6 o'clock. Do you think she is talking about 6 o'clock in the morning or 6 o'clock in the evening? Why?

# Unit 1 Time of Day

## Let's Think ...

This clock shows time on the hour at 7 o'clock.

- How many different on the hour times can this clock show?
- There are 24 hours in a day. How does the clock manage to show all these times?

Each day we experience **morning**, **afternoon**, **evening** and **night**. A day is 24 hours long and, when it is over, a new day starts.

The clock above only shows 12 hours. This means that each time happens twice in one day; for example, the clock shows 6 o'clock in the morning and it shows 6 o'clock again in the evening.

We can write the letters **a.m.** and **p.m.** to show whether the time is in the morning or in the afternoon.

Times from **midnight** until just before **midday** (**noon**) are a.m. times.

Times from noon until just before midnight are p.m. times.

3 o'clock in the morning is 3:00 a.m.
3 o'clock in the afternoon is 3:00 p.m.

1 Match each picture with the time it might happen.

   a 9:00 a.m.     b 11:00 a.m.     c 3:00 p.m.     d 6:00 p.m.

**2** Read the time on each clock. Say one thing that happens at that a.m. time and one thing that happens at that p.m. time.

a     b     c     d

**3** Use the times on the clocks to complete each sentence. Remember to include a.m. or p.m. when you say the time.

a Soccer practice starts at  and ends at  .

b The evening news starts at  and ends at  .

c Kayleigh went to the beach at  and

went home at  .

d Mr Jonas went to bed at  , but the noise of a storm woke

him up at ◔ .

**Looking Back**

Say whether each event happens at an a.m. time or a p.m. time.

eat dinner     morning     wake up     go to bed     sunrise

evening     eat breakfast     sunset     afternoon     go to school

# Unit 2  Telling the Time

**Let's Think …**

- What does the long hand on a clock show?
- What does the short hand on a clock show?
- How many minutes are there in an hour?
- How many minutes are there in a half hour?
- What number is the long hand on when it is half-past an hour?
- Is the short hand on 12 when it is an o'clock time?
- How many half hours are the same as one hour?

*You already know how to tell time on the hour and half hour.*

*This clock shows 10 o'clock.*

*The long hand is on the 12. The short hand shows the hour.*

*This clock shows **half-past** 10.*

*The long hand is on the 6. It has moved halfway round the clock.*

*The short hand is halfway between the 10 and 11.*

*We can also tell time to the quarter hour.*

*This clock shows **quarter-past** 10.*

*The long hand is on the 3. It has moved one quarter $\left(\frac{1}{4}\right)$ of the way round the clock.*

*The short hand is past the 10, but not near the 11 yet.*

*This clock shows quarter-to 11.*

*The long hand is on the 9. It has only got one quarter $\left(\frac{1}{4}\right)$ of the way to go to get back to the hour.*

*The short hand is nearly on the 11.*

**1 What time is shown on each clock?**

a    b    c    d

**2 Choose the correct time.**

a  quarter-past 9 or quarter-to 11

b  quarter-past 5 or quarter-to 6

**3 These clocks show p.m. times.**

a Read the time on each clock.

b Which time is earliest in the day?

c Which time is latest in the day?

d Which time is closest to 4 o'clock?

e What is the time a quarter of an hour after the time shown on each clock?

4 Marie catches the bus at 6 o'clock. The bus then takes quarter of an hour to get to her stop. What time does she arrive at her stop?

**Looking Back**

Where is the long hand of the clock at each time?

a half-past seven       b quarter-to seven       c quarter-past seven

# Topic Review

**Talking Mathematics**

Choose the correct labels.

| on the hour | half-past | quarter-to | quarter-past | long hand | short hand | 9 o'clock |
|---|---|---|---|---|---|---|

**Quick Check**

1 These clocks all show a.m. times.

   a What time is shown on each clock face?

   b What might you be doing at each time?

2 What is the time:

   a half an hour after half-past three    b quarter of an hour after half-past three?

3 Add these amounts in your head.

   a 15 + 10    b 18 + 10    c 23 + 10    d 87 + 10    e 10 + 55    f 30 + 50

4 A shape has four sides. Two of the sides are long and two of the sides are short. All the corners are the same size. What shape is it?

# Topic 20 Introducing Multiplication

Workbook pages 61–63

▲ Each window on the house has three opening panes. Can you work out how many panes there are without counting each one?

You already know how to **count** in groups by **skip-counting**. When you skip-count by 5s, you count 5, 10, 15, 20 and so on. If you picture hops on a number line, you will see that skip-counting by 5 is the same as **adding** 5 each time. Counting 5, 10, 15, 20 is the same as 5 + 5 + 5 + 5. In this topic, you will learn to use the idea of repeated addition to add groups of numbers quickly.

## Getting Started

1 Use skip-counting to answer these questions.

a How many ears altogether in your group?

b How many fingers altogether in your group?

c How many hands and feet altogether in your group?

2 How can you quickly work out how many sides there are altogether in this group of triangles?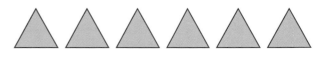

# Unit 1 Repeated Addition

There are 2 fish in each row.

We can add on 2 each time to find how many there are:

2 + 2 + 2 + 2 = 8

This is called **repeated** addition.

2 + 2 + 2 + 2 is four sets of two. We have added 2 four times.

This is the same as skip-counting by 2s for four counts: 2, 4, 6, 8

**Multiplication** is another name for repeated addition. It is a quick way of adding groups to get a total.

We use the sign × to show how many times we have added each number.

2 + 2 + 2 + 2 = 8 can be written as 4 × 2 = 8

We read this as 4 times 2 equals 8.

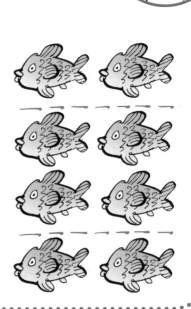

---

1  **What is the total of each set of items?**

a

2 + 2 + 2 + 2 + 2 + 2 = ☐

b

4 + 4 + 4 = ☐

c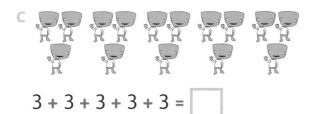

3 + 3 + 3 + 3 + 3 = ☐

d ▭▭▭▭ ▭▭▭▭ ▭▭▭▭ ▭▭▭▭

10 + 10 + 10 + 10 = ☐

2 For each set of dots:

- count the number of rows
- count the number of dots in each row
- write the repeated addition sentence and the multiplication sentence.

a • • • • •
• • • • •
• • • • •
• • • • •

b • • •
• • •
• • •
• • •
• • •

c • • • • • • •
• • • • • • •

d • • • •
• • • •
• • • •
• • • •
• • • •

e • • • • • • • • • • •
• • • • • • • • • • •
• • • • • • • • • • •
• • • • • • • • • • •
• • • • • • • • • • •

f • • • • •
• • • • •
• • • • •
• • • • •
• • • • •
• • • • •
• • • • •

## Looking Back

1 A scooter has 2 wheels. How many wheels are there on 8 scooters?

2 A crab has 10 legs. There are 7 crabs in a rock pool. How many legs are there altogether?

# Unit 2 More Multiplication

**Let's Think ...**

- How many points altogether on these stars?
- How could you write this as a repeated addition sentence?
- How could you write this as a multiplication sentence?

*Remember, multiplication is the same as repeated addition.*

*3 + 3 + 3 + 3 + 3 =* 15

*5 groups of 3 =* 15

*5 × 3 =* 15

1 Write a repeated addition sentence and a multiplication sentence for each part.

a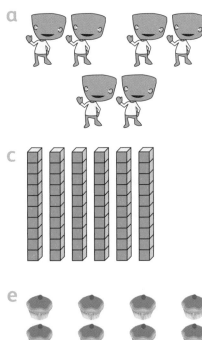

b
★ ★ ★ ★ ★
★ ★ ★ ★ ★
★ ★ ★ ★ ★

c

d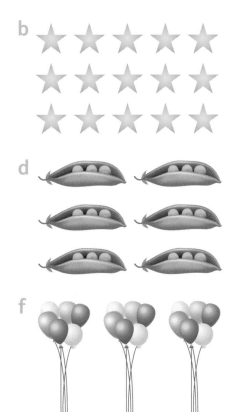

e

f

**2** How many legs?

a
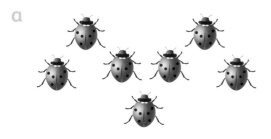

b

**3** There are seven goats in a field.

a How many eyes altogether?　　b How many legs in all?

**4** Michael sells beaded bracelets to tourists. He uses 10 beads to make one bracelet.

a How many beads will he need to make 12 bracelets?

b He sells 8 bracelets for $5.00 each. How much money is this?

**5** Count the number of eyes each alien has.

A 　　B 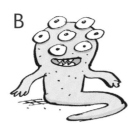　　C 　　D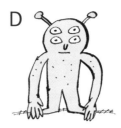

a 8 of alien A are found in a cave. How many eyes is this?

b 10 of alien B are found underwater. How many eyes altogether?

c 9 of alien C are found in a mango tree. How many eyes are there?

d 10 of alien D are found in a shoe. How many eyes is this?

---

**Looking Back**
Draw dot pictures to show each number sentence.
a 3 + 3 + 3　　　　　b 10 + 10 + 10
c 4 × 2　　　　　　　d 5 × 4

# Topic Review

**What Did You Learn?**

- We can use repeated addition to count groups of objects quickly.
- Repeated addition is also called multiplication.

**Talking Mathematics**

Read these sentences and say the missing words.

- The picture shows ____ groups of ____ dots.

- We can add ____ each time to find the total score on the dice.

- Adding 3 over and over is called ____.

- 3 + 3 + 3 + 3 + 3 + 3 = ____

- Six times three is ____.

**Quick Check**

1 Write a repeated addition sentence and a multiplication sentence for each picture.

a ★ ★ ★
  ★ ★ ★
  ★ ★ ★
  ★ ★ ★

b, c, d

2 Show how you could work out the score on each dartboard.

a

b

c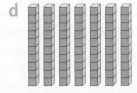

3 Leeandra wants to arrange 30 dots in rows. Draw pictures to show how she could arrange them to have:

a 2 dots in each row  b 3 dots in each row  c 6 dots in each row  d 3 × 10 dots.

# Topic 21  Measuring Mass

Workbook
pages 64–65

▲ What can you see in this photograph? What do you call the
measuring instrument? What is being measured?

Last year you learned to describe items as heavy or light and compared
different items to find out which was **heavier** or **lighter**. In this topic, you
are going to learn about the instruments and units for measuring how
heavy or light something is.

## Getting Started

1 Which item in each pair do you think is heavier?

  a 5 pencils or 5 erasers

  b a story book or a copy book

  c 8 blocks or 8 beads

  d 2 markers or 2 crayons

  e an apple or a calculator

2 How can you check?

# Unit 1 Kilograms

**Let's Think ...**

- There are three children on both sides of the see-saw.
- Why is the see-saw not balanced?
- Which group of children is lighter? How do you know?

*We often use the word **weight** when we talk about how heavy or light things are. The correct word for this in mathematics is **mass**. We can measure the mass of objects in a **metric** unit called **kilograms**. We can write kilograms in short form as kg.*

*Look at these four **balance scales**.*

*The **pans** are level so we know that the objects in the left and right pans have the same mass. We say the scales are **balanced**.*

a      b      c      d

1 Look at these sea creatures.

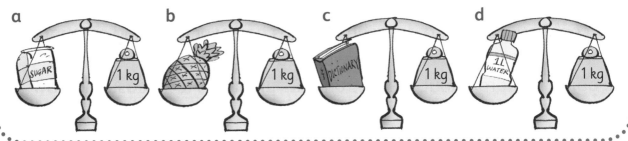

  a Which creature is the heaviest?

  b Which creature is lighter, the lobster or the seahorse?

  c Write the names of the creatures in order of their mass, starting with the lightest.

**d** Next to the name of each creature, write down whether you think it has a mass of more than 1 kilogram (> 1 kg), about 1 kilogram (1 kg) or less than 1 kilogram (< 1 kg).

**e** An adult man has a mass of about 85 kg. Esimate the mass of a shark.

**2** Look at the balance scales carefully.

Which items have a mass of:

**a** exactly 1 kg

**b** more than 1 kg (> 1kg)

**c** less than 1 kg (< 1kg)?

**3** Work in groups. Your teacher will give you a balance scale and a kilogram weight.

| My items | My estimate | My measurement |
| --- | --- | --- |
| | | |
| | | |

**a** Draw a table like this one in your copy book.

**b** Write the names of five items.

**c** Estimate whether each item will be **<** 1 kg, about 1 kg or **>** 1 kg. Write your estimate in the second column.

**d** Use the balance scale and kilogram weight to measure the mass of each item. If your estimate was correct, tick the third column. If it was not correct, write the correct statement in the third column.

---

## Looking Back

**1** A farmer loaded a 29 kg sack of sugar and a 12 kg hand of bananas onto his truck. What is the total mass of these items?

**2** Mrs Gleeson harvested 83 kg of cabbages. She sold 49 kg of them at the market. How many kilograms of cabbages did she have left?

# Unit 2  Kilograms and Grams

## Let's Think …

Marsha buys two bags of fruit and puts them on her balance scale.

- What is the mass of each bag of fruit?
- Complete this sentence.
  One kilogram is the same as ☐ grams.

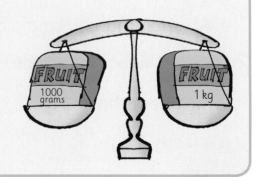

> We use **grams** to measure objects that are lighter than a **kilogram**.
> 1 kilogram is the same mass as 1000 grams.
>
> $1 \text{ kg} = 1000 \text{ g}$     $\frac{1}{2} \text{ kg} = 500 \text{ g}$     $\frac{1}{4} \text{ kg} = 250 \text{ g}$

**1** Arrange these masses in order from lightest to heaviest. Write each mass in short form using numbers and the symbol g.

fifty grams    two hundred fifty grams    twenty-five grams

eight hundred grams    one thousand grams    five hundred grams

forty grams    seventy-five grams    one hundred ninety grams

**2** The mass of each item is shown in grams.

Work out the mass of:

a the lunch and the juice

b the crackers and the sweets

c the water and the juice

d the lunch and the crackers.

**3** Which three items in the picture in Question 2 have a total mass of 1 kg?

**4** Mario eats half the crackers shown in the picture in Question 2. What mass of crackers is this?

**5** Choose the correct mass for each item.

a       about 140 g     or     about 14 kg

b       about 100 g     or     about 1 kg

c       about 20 g     or     about 20 kg

d       about 2 g     or     about 2 kg

e       about 100 g     or     about 1 kg

f       about 100 g     or     about 1 kg

**Looking Back**

**1** How many 500 g weights are needed to balance 1 kg?

**2** Sarah buys 1 kg of sugar. She divides it equally between 4 smaller packets.

   **a** What is the mass of each small packet in kilograms?

   **b** What is the mass of each small packet in grams?

# Topic Review

**What Did You Learn?**

- Mass is a measure of how light or heavy something is.
- We can measure mass in kilograms or grams.
- There are 1000 grams in 1 kilogram.

**Talking Mathematics**
Describe each picture using the words you have learned about mass.

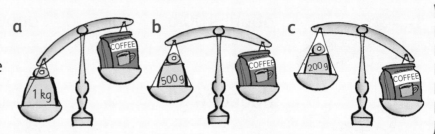

**Quick Check**

1 Order these objects from lightest to heaviest. Estimate the mass of each item.

2 Each scale is balanced. Write down the mass of the object in the left pan.

  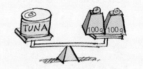 

3 Bananas cost $3.00 per kg. How much would you pay for:

a 4 kg          b 10 kg?

4 A toy car weighs 250 g. There are 4 cars in a pack. Is the mass of the pack < 1kg, 1 kg or > 1 kg?

5 Jessie's lunch weighs 300 g. Michaela's lunch weighs half this amount. What is the mass of Michaela's lunch?

6 The mass of half a sack of pineapples is 12 kg. What is the mass of a whole sack of pineapples?

# Topic 22  Solid Shapes Workbook pages 66–68

▲ Find the solid shapes in this picture. Do you remember the correct name for each shape?

## Key Words
solid
cube
rectangular prism
sphere
cone
cylinder
pyramid
face
attributes
stack
slide
roll

Lots of the real objects in the picture are solid shapes with mathematical names. For example, a ball is a **sphere** and the ice-cream cone is the shape of a **cone**! In this topic, you are going to revise the names and properties of solid shapes. You will also sort and classify solids according to their attributes.

## Getting Started

1 Think of as many objects as you can that are shaped like a sphere.

2 Food cans are shaped like cylinders. What other things are shaped like a cylinder?

3 Look around the classroom. Which solid shape is most common?

# Unit 1  Shapes and their Properties

## Let's Think ...

● Which solid shape is each child thinking of?

● How do you know?

> It is shaped like a box.

> It is shaped like a beach ball.

> It has a flat bottom and triangular faces that meet at a point.

---

*Look at these **solid** shapes.*

cube    sphere    rectangular prism    pyramid    cylinder    cone

*The flat surfaces of solid shapes are called **faces**.*

*We can use the number of faces a shape has to classify it.*

*A **cube** has six square faces. They are all the same size.*

*A **rectangular prism** has six rectangular faces. The faces opposite each other are the same size.*

*A **sphere** has no flat faces. Its surface is curved.*

*A **cylinder** has two circular faces.*

*A **cone** has one circular face.*

*A **pyramid** has one face at the bottom and triangular faces that meet at a point. The number of faces depends on the base.*

1 Name each shape.

  a A solid shape with six square faces.

  b A solid shaped like a can.

  c A solid, pointed shape with a circular base.

  d A round object with one curved surface.

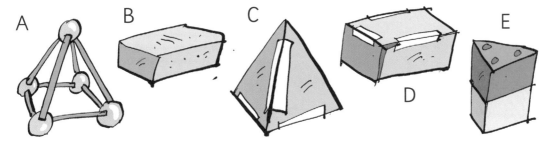

**2** The pictures show models of solid shapes made by a group of Class 2 students.

A    B   C    D   E

**a** Name each shape.

**b** Describe how the students made each model.

**3** Work in groups. Make your own models of a cube, a rectangular prism, a pyramid and a sphere. Decide how you will make the models before you start.

**4** Tarez joined some models together to make these objects:

A      B      C

**a** What shapes did she use to make each object?

**b** Draw or make objects by joining these pairs of shapes:

- a cube and a rectangular prism
- two cones
- two rectangular prisms
- a cone and a sphere.

**Looking Back**
What shape is each object?

# Unit 2  Sort and Classify Shapes

## Let's Think ...

Can this cone:

- roll
- slide
- stack?

Use a model to check.

We can **sort** shapes into groups or **classify** them using their properties or **attributes**.

We can sort shapes by the number of faces they have.

| No faces | 1 face | >1 face |
|----------|--------|---------|
|          |        |         |

We can also sort shapes by whether they can **stack**, **roll** or **slide**.

We say that shapes stack if they can be arranged on top of each other.

We say that shapes roll if they can move by turning over.

We say that shapes can slide if they can move across a surface without turning over.

**1** Say whether these objects can stack, roll or slide.

a    b    c    d    e

f    g    h    i    j

2 Look around the classroom. How many solid shapes can you find? List the objects that can:

a stack

b slide

c roll

d roll and slide

e roll and stack

f stack, slide and roll.

3 Elisha built this model from solid wooden shapes.

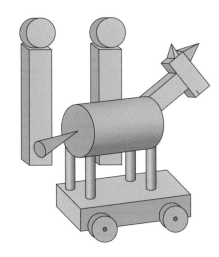

a Write down the names of the solids he used.

b How many of each solid shape did he use?

**Looking Back**

Tamika sorted and classified a set of shapes like this.

| Group 1 | Group 2 | Group 3 |
|---------|---------|---------|

a What attributes did she use to sort the shapes?

b What name could you give each group of shapes?

# Topic Review

## Talking Mathematics

Match each word with the correct meaning.

| Word | Meaning |
|------|---------|
| Face | Place into groups. |
| Roll | Flat surface of a solid. |
| Slide | Move by turning over. |
| Classify | Move across a surface without turning over. |

## Quick Check

1 Which solid shapes have:

   a only flat surfaces   b only curved surfaces   c both flat and curved surfaces?

2 Name a solid that can:

   a roll and stack       b roll but not stack       c slide, roll and stack.

3 Which solid shape is missing from this pattern?

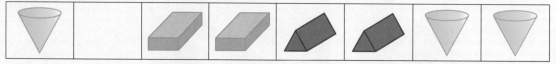

4 A cube has six faces. Write a repeated addition sentence to find the number of faces on six cubes.

5 Copy each statement. Write < or > in the boxes to make the statements true.

   a 456 ☐ 564   b 546 ☐ 465   c 654 ☐ 645

6 Draw a repeating pattern using three different solid shapes.

# Topic 23 Area Workbook pages 69–71

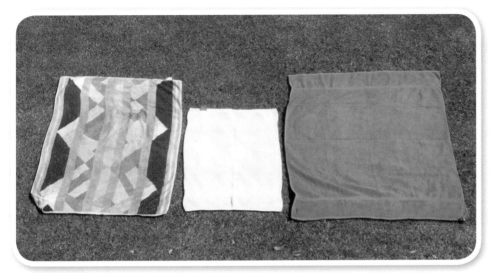

## Key Words
surface
area
covered
equal
square unit

▲ Which towel covers the most grass? Which towel covers the least grass? How do you know?

Last year, you covered the surface of shapes with tiles and other objects to measure area. In this topic, you are going to use squared grids to work out the area of different shapes.

## Getting Started

This map is drawn on a square grid. The area of island A is about 7 squares.

1 Estimate the area of each of the other islands.

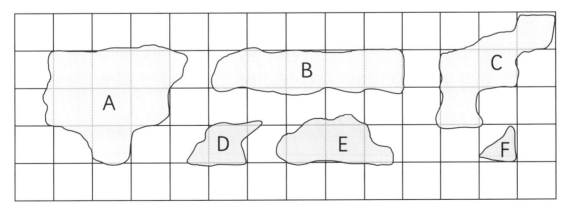

2 Will a town with an area of 3 squares fit on island C?

3 Estimate the area of the map that is covered with water.

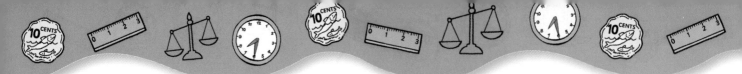

# Unit 1  Area in Square Units

Let's Think …

Look at these sticky notes.

- Which note will take up most space on a page?

- Romeo says that notes A and D have the same area. How could you check this?

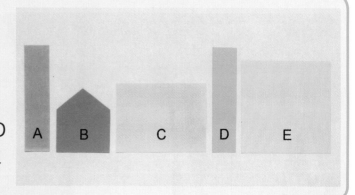

Area is the amount of **surface covered** by a flat shape.

*You can find the area of a shape drawn on a grid of **equal** squares by counting the number of squares the shape covers.*

*Area is measured in **square units**.*

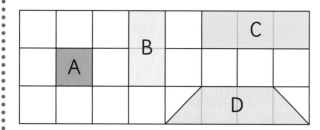

- *Shape A has an area of 1 square unit.*

- *Shape B has an area of 2 square units.*

- *Shapes C and D have the same area – 3 square units.*

1 Work out the area of each shape. Write down the name of each shape and its area.

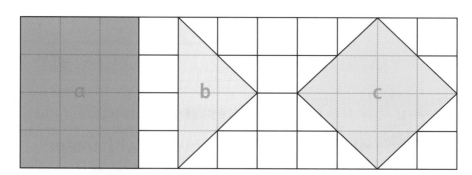

**2** Shamar and Xavier are lost. They spell out this word on the ground using square tiles.

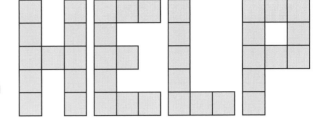

  **a** What is the area of each letter in square units?

  **b** How many square tiles did they use in all?

**3** Shandra says these shapes all have the same area. Is she correct? How do you know?

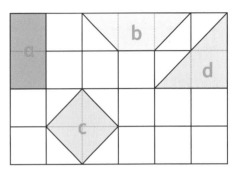

**4** Curtis put some leaves on a square grid to estimate their area, but he could not see the squares through the leaves. Try to estimate the area of each leaf.

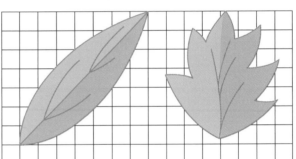

**5** Your teacher will give you a grid of squared paper and some small leaves. Use the grid to estimate the area of each leaf. Then paste the leaves onto the sheet and write the area next to each one.

---

## Looking Back

**1** Estimate the area of this shape.

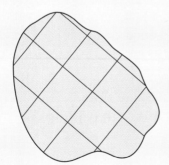

**2** Work out the area of this shape.

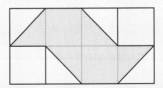

# Topic Review

**Talking Mathematics**

Mosaic patterns are often made from small square tiles. They can be used to make patterns on walls and floors. The picture shows part of a much larger mosaic pattern from the floor of a palace near Jericho. It is over 1300 years old.

- How can you find the area of each colour in the pattern?
- Find the area of each colour in the pattern. Work with a partner.

**Quick Check**

1 What is the area of each letter in this word?

2 Why did you only need to find the area of three of the letters?

3 The areas of some pairs of shapes are given below. Work out the total area of each pair.

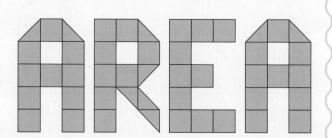

a 32 square units + 64 square units

b 33 square units + 44 square units

c 70 square units + 27 square units

d 31 square units + 27 square units

4 Misha has a square with an area of 64 square units. He cuts a smaller square with an area of 25 square units out of the middle. What is the area of the shape that is left?

# Topic 24  Introducing Division

Workbook pages 72–74

### Key Words

multiplication

division

divide

share

equal groups

repeated subtraction

▲ Skip-count by threes to find the number of eggs in the tray. How many groups of three are there? How many groups of two can you make?

You already know how to use repeated addition to combine equal groups and find the total. In this topic, you are going to start with the total and make smaller groups. You will be dividing a larger amount into smaller equal amounts.

## Getting Started

Zarea has 18 pennies.

Use counters to help you work out the answers to the questions.

1 How many groups of 3 can she make?

2 How many groups of 6 can she make?

3 How many groups of 2 can she make?

4 Can she make equal groups of 5 with this number of pennies? Explain your answer.

# Unit 1 Division as Repeated Subtraction

**Let's Think ...**

- How many times can you subtract 3 from 21?
- Use the number line to help you.

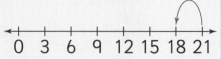

0  3  6  9  12 15 18 21

---

*You already know that **multiplication** is repeated addition.*

*For example: 7 × 3 = 3 + 3 + 3 + 3 + 3 + 3 + 3 = 21*

**Division** *is repeated subtraction. When you divide, you share the total equally between a number of groups. To divide, you start with the total and then subtract **equal groups** until you reach 0.*

*For example:* 21 $\underset{①}{-3}$ $\underset{②}{-3}$ $\underset{③}{-3}$ $\underset{④}{-3}$ $\underset{⑤}{-3}$ $\underset{⑥}{-3}$ $\underset{⑦}{-3}$ = 0

- *We subtracted 3 seven times.*

- *There are 7 groups of 3 in 21.*

---

1 Find the answers to these division problems by subtracting equal groups.
You can use counters to model the problems.

a How many groups of 2 are there in 24?

b How many groups of 3 are there in 24?

c How many groups of 4 can you make with 24?

d How many groups of 6 can you make with 24?

e How many times can you subtract 8 from 24?

**2** There are 20 mugs on a shelf.

    a How many groups of 10?

    b How many groups of 4?

    c I arrange the mugs in sets of 2. How many sets are there?

**3** Mrs Pinter has these push pins in her classroom.

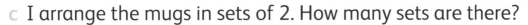

    a Skip-count by 3s to find out how many pins she has in all.

    b She uses 3 push pins to put up one picture. How many pictures can she put up?

    c She hands out 10 push pins to each group in the class. How many groups are there?

    d How many groups of 6 can she make?

    e What is 30 divided by 5?

**4** Sondra has 10 popsicles. She eats 2 a day. How many days will they last?

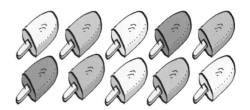

**5** A sports club has 16 cricket bats and 14 cricket balls.

    a How many sets of 4 bats can they make?

    b Each team needs 2 balls. How many teams can get 2 balls?

**6** Suresh puts a counter on the number 27 on a number line. He makes backwards hops of 3 until he gets to 0. How many backwards hops is this?

**7** A spider has 8 legs. Nadia counts 24 spider legs. How many spiders are there?

---

**Looking Back**

**1** How many times can you subtract 2 from 16?

**2** How many equal groups of 4 can you make with 24?

# Topic Review

**Talking Mathematics**

Curtis made this chart to help him remember multiplication words.

- Make a card to help you remember division words. Work with a partner.
- Discuss which words to include.
- Make sure you understand what each word means.

Multiplication

multiply, lots of, times, groups of, multiplied by and repeated addition

**Quick check**

1 Here are 20 counters.

  a How many groups of 2 can you make?
  b How many times can you subtract 4 from 20?
  c What is 20 divided by 2?

2 Chelsea has one hundred 1¢ coins. She arranges them in piles of 10. How many piles can she make? Write a repeated subtraction for this story.

3 A tin holds 5 litres of paint. How many tins can Mr Jones fill from a 25 litre barrel?

# Topic 25 Probability Workbook pages 75–76

▲ Talk about this photograph. Do you think it is possible for an elephant to surf a wave like this? Why or why not?

Last year, you learned that probability is about how likely something is to happen. In this topic, you are going to learn about things that happen in daily life and decide whether or not they are likely.

## Getting Started

1 Which of these events have a good chance of happening?

2 Which have no chance of happening? Why?

Our principal will come to school on Monday wearing a bathing suit.

It will be warm tomorrow.

If I drop my pencil, the lead will break.

If I roll a dice, I will get a 7.

Our next teacher will be a robot.

# Unit 1 Likely and Unlikely Events

**Let's Think ...**

If you take out a ball from each bowl without looking, which colour ball do you think you will get? Why?

a          b          c

*We use special words to describe the **probability** that something will or will not happen.*

*Things that will definitely happen are **certain**.*

*It is certain that the sun will rise tomorrow.*

*Things that are **likely** have a good chance of happening.*

*It is likely that you will roll a number greater than 1 when you roll a dice.*

*Things that are **unlikely** have a small chance of happening, but they are **possible**.*

*It is unlikely that you will win a car in a competition.*

*Things that cannot happen are **impossible**.*

*It is impossible to hold your breath for a week.*

1  Josiah is rolling a normal dice. Say whether these things are *certain, likely, unlikely* or *impossible*.

  **a** He will roll a 6 five times in a row.

  **b** He will roll a zero.

  **c** He will roll an odd number.

  **d** He will roll a number between 1 and 6.

**2** Look at these three spinners.

A 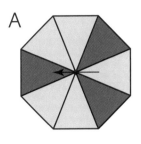  B 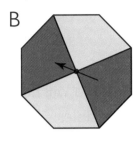  C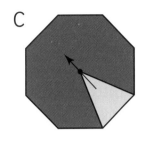

    **a** Which spinner is most likely to stop on yellow?

    **b** Which spinner is most likely to stop on red?

    **c** Which spinner is least likely to stop on yellow?

    **d** Which spinner has a fair chance of stopping on red or yellow? Why?

**3** List three things that are likely to happen to you today.

**4** List three things that are unlikely to happen to you today.

**5** Zara has these pencils in a bag to hand out to the students in her class. She takes the pencils out of the bag without looking.

    **a** Which colour pencil is she most likely to give the first student? Why?

    **b** Is the yellow pencil *likely* or *unlikely* to be chosen first? Why?

    **c** The first pencil she gives out is yellow. What colour pencil is the next student likely to get?

---

**Looking Back**

Look at these spinners.

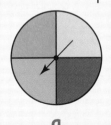 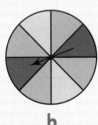 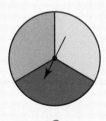 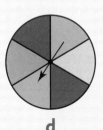 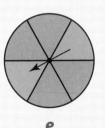
   a      b      c      d      e

Choose the probability word that describes the chance that each spinner stops on blue.

# Topic Review

**Talking Mathematics**

Zion has mixed up some probability words. Correct his mistakes.

- It is *certain* that it will snow today in Nassau.
- It is very *likely* that our teacher will cut fruit with an axe.
- It is *unlikely* that we will have homework this week.
- The sun is *impossible* to set tonight.
- When you roll a dice, it is *likely* that you will get a 7.

**Quick check**

1 Jessie put 15 red counters, 10 yellow counters and 5 green counters into a bag.
   a How many counters is this altogether?
   b If he takes out a counter without looking, what colour is it likely to be?
   c Which colour is he least likely to pick? Why?

2 Draw a spinner with three colours that is:
   a likely to land on yellow
   b unlikely to land on green
   c certain to land on yellow, green or blue.

3 What is the probability?
   a It will rain and be sunny at the same time.
   b You will go to bed tonight.
   c There will be a flood at school today.
   d You will drive yourself to school.

4 Complete these sentences.
   a There are ☐ groups of 4 in 20.
   b There are 2 groups of ☐ in 20.
   c Half of 20 is ☐.

5 I paid for an item costing $1.05 with $2.00. My change was five coins. What coins could they be?

# Topic 26 Exploring Patterns and Shapes
Workbook pages 77–78

## Key Words
pattern
natural
plane shape
solid
two-dimensional
three-dimensional

▲ What patterns can you see in this picture? Where else can you see natural patterns?

We can find natural patterns and patterns made by people all around us. In this topic, you are going to explore patterns and learn how you can use solid shapes to make patterns of flat shapes.

## Getting Started

1 Can you see any patterns in your classroom? Look at the floor, walls, tables and windows. Describe the patterns you see.

2 Which patterns are made from shapes?

3 Patrice put paint on the faces of a solid shape and used them to stamp this pattern. What solid do you think he used?

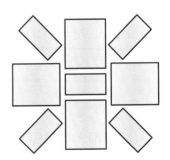

# Unit 1  Patterns Everywhere

**Let's Think …**

These pictures show parts of patterns. Where does each one come from?

*We can see **patterns** all around us.*

*Some patterns are **natural**. Fish scales and veins on leaves are examples of natural patterns.*

*Some patterns are made by people; for example, tile patterns, blockwork patterns and fabric designs.*

Work in a group.

**1** Find at least 10 patterns in your environment that you like.

**2** Make a poster showing these patterns.

Give your poster a title.

Display your patterns. You can draw them, take photographs or find pictures of similar patterns.

**3** For each pattern, write where you found it and why you chose it.

**4** Show your poster to the class and tell them about the patterns you chose.

**Looking Back**

**a** Design a pattern for a rug that is based on a natural object.
**b** Draw your pattern in your copy book.

# Unit 2 Flat Shapes and Solids

**Let's Think …**

What flat shapes can you see on these solids?

*Flat shapes are **two-dimensional** (2-D). They have length and breadth.*
*Solid shapes are **three-dimensional** (3-D). They have length, breadth and height.*
*The faces of three-dimensional **solids** are often shaped like **plane** shapes.*
*For example, each face of a cube is a square.*

1 Jamila has 3-D blocks like the ones in the picture. She uses them to print shape patterns onto fabric.

A    B    C    D    E    F

a Which solids can she use to stamp circle shapes?

b Which solids can she use to stamp square shapes?

c Which solid can she use to stamp rectangular shapes?

d Can she use any of these solids to stamp triangular shapes?

2 You push solid shapes through the holes in this children's toy.

Look at the 3-D blocks in Question 1.

Which hole could you post each block through? Some may fit through more than one hole.

**Looking Back**

1 Which shapes have faces shaped like a circle?

2 Which shapes can have square faces?

# Topic Review

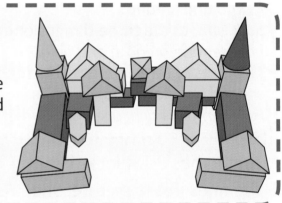

## What Did You Learn?

- Patterns are found all around us. Some are natural, some are made by people.
- Flat shapes are two-dimensional.
- Solid shapes are three-dimensional. Their faces are often flat shapes.

## Talking Mathematics

Some Grade 2 students built this model.

a With a partner, describe this model. Use the correct mathematical words for the 2-D and 3-D shapes.

b What patterns can you see in the model?

## Quick Check

1 Read the descriptions. Name each shape and say whether it is two-dimensional or three-dimensional.

  a It has 4 equal sides.

  b It has 6 square faces.

  c It has a curved surface and no faces.

  d It has a curved surface and one round face.

  e It has 3 corners.

  f It is pointed and has some triangular faces.

  g It has 4 corners and 4 equal sides.

2 Choose *solid* or *plane* to complete each sentence correctly.

  a A sphere is a ___ shape.

  b A cylinder is a ___ shape.

  c A triangle is a ___ shape.

  d A cube is a ___ shape.

  e A cone is a ___ shape.

3 Use mathematical words to describe the shape of:

  a a basketball

  b a soup can

  c a sheet of paper

  d the face of a round clock

  e a building block.

# Topic 27 Looking Back

## Revision A

1 Write the numeral for:

   a four hundred twenty

   b six hundred eighty-three

   c two hundred four.

2 What is the difference between 17 and 9?

3 Copy and complete the number sentences.

$9 + 5 = \boxed{\phantom{0}}$     $14 - 5 = \boxed{\phantom{0}}$

$5 + \boxed{\phantom{0}} = 14$     $14 - \boxed{\phantom{0}} = 5$

4 John has 3 marbles. He wants 20 marbles. How many more marbles does he need?

5

   a What are the next three shapes in this pattern?

   b What is the third shape in the pattern?

   c In which positions are the circles?

6 Write the month:

   a after May

   b before September.

7 Write the numbers shown in numerals and in words.

a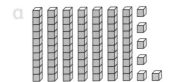

b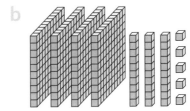

8 Draw a grid like this in your copy book.

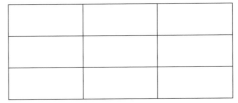

Draw a star in the box in the middle.

Tick the box above the star.

Draw a heart in the box left of the tick.

Write one hundred three in numerals in the box beneath the star.

→

Write the number one less than one hundred three in numerals in the box to the right of the star.

Draw a circle in the box to the left of the star.

9 Complete these sentences.

  a Today is _____.

  b Tomorrow will be _____.

  c Yesterday was _____.

10 Copy and complete the table.

| One less | Number | One more |
|----------|--------|----------|
|          | 431    |          |
|          | 500    |          |

11 What is:

  a in the tree

  b under the tree

  c next to the tree

  d in front of the swing?

12 a What do we use to measure temperature?

  b Is a temperature of 20 degrees Fahrenheit hot or cold?

# Revision B

1 Janelle threw two darts at this board. She adds together the numbers they land on to get her score. Write number sentences to show all the scores she could get.

2 Draw a square.

  a How many sides does it have?

  b How many sides are there on two squares?

3 How many dimes have the same value as 50¢?

4 Write the name of each shape.

  a        b        c

5 How much money?

  a

  b

c

d
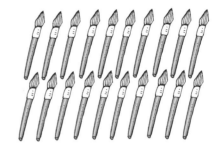

6 Draw two circles.

  a Shade half of the first circle.

  b Divide the second circle into fourths.

7 Write the number that is 100 more than:

  a 200      b 450      c 875.

8 Draw a picture to show how you could make $2.00 using exactly five coins.

9 What fraction of each shape is shaded?

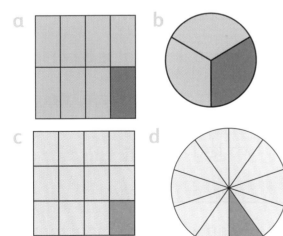

10 Count the brushes.

  a How many in half the group?

  b How many in $\frac{1}{4}$ of the group?

  c I pick up $\frac{1}{10}$ of the brushes. How many is this?

11 Estimate and then count the number of stars.

12 a Count the number of chairs and tables in your classroom.

  b Draw this frame into your copy book. Complete the frame to draw a pictograph showing your information.

| Furniture in our class | |
| --- | --- |
| Chairs | |
| Tables | |

13 a 23 + 34      b 19 + 26

  c 42 − 20      d 63 − 26

# Revision C

1 Answer these questions about the calendar.

**JUNE**

| S | M | T | W | T | F | S |
|---|---|---|---|---|---|---|
| 1 | 2 | 3 | 4 | 5 | 6 | 7 |
| 8 | 9 | 10 | 11 | 12 | 13 | 14 |
| 15 | 16 | 17 | 18 | 19 | 20 | 21 |
| 22 | 23 | 24 | 25 | 26 | 27 | 28 |
| 29 | 30 | | | | | |

a How many days in June?

b How many Mondays in June?

c What day is June 12th?

d What month comes after June?

e Peter's birthday is on June 14th. Is this on a weekend?

2 Copy and complete these number sentences.

□ + □ + □ + □ = □

□ groups of □ = □

□ × □ = □

3 Micah has these coins.

a Write a repeated addition sentence to show the total.

b Write a multiplication sentence to show the total.

4 Here are pictures of four objects.

Which object is:

a about 1 m wide

b about 1 dm tall

c about 15 cm long

d about 2 m high

e about 3 cm wide?

5 Draw a line that is 8 cm long.

6 How many sides are there on six triangles? Show how you worked this out.

7 *Likely* or *unlikely*?

a You will have school tomorrow.

b It will be warm today.

c Your teacher will fly to school.

160

**8** What time is shown on each clock?

a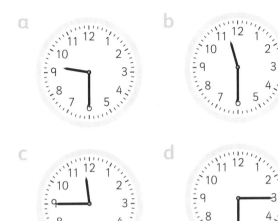
b

c
d

**9** Is the red line on each shape a line of symmetry?
Write yes or no.

a
b

c
d

**10** Which of these shapes are symmetrical?

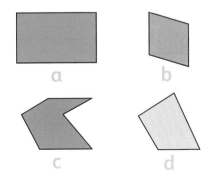

a

b

c

d

# Revision D

**1** How many groups of 4 can you make from:

  a 8 eggs    b 16 socks

  c 20 chairs?

**2** What is 15 divided by 5?

**3** How many grams are there in:

  a 1 kilogram  b $\frac{1}{2}$ kilogram?

**4** Which is the heaviest mass in each group?

  a 372 g, 723 g, 237 g

  b $\frac{1}{2}$ kg, 580 g, 2 kg

**5** Look at these shapes.

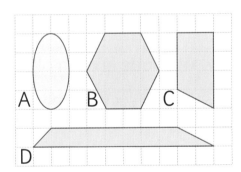

Which shape has:

  a the greatest area

  b the smallest area

  c an area smaller than shape C

  d an area greater than shape D?

**6** There are 20 sweets in a packet. $\frac{1}{2}$ are cherry, $\frac{1}{4}$ are banana, one is lime and the rest are apple.

How many are:

a cherry    b banana    c apple?

**7** How many days in:

a December    c March and April

b May         d 5 weeks?

**8** Which of these numbers have 90 as the closest ten?

88, 96, 91, 99, 94

**9** Look at these everyday objects.

a What shape is each object?

b Which object has only flat faces?

c Which solid has no faces?

d Which solid cannot roll?

e Which object cannot slide?

f Which two objects can stack and slide?

g Draw the faces of a rectangular prism.

**10** Jasmine sees 20 boats in the harbour.

a How many groups of 2 could there be?

b How many groups of 4 could there be?

c How many groups of 5 could there be?

d How many groups of 10 could there be?

e The next day there are half as many. How many is this?

f What is double 20?

**11** One box can hold 10 mangos.

a How many boxes are needed to hold 50 mangos?

b How many mangos are there in 4 boxes?

c Mike has 3 boxes of mangos. Half of one box are rotten. How many are not rotten?

**12** Write one thing that is certain to happen tomorrow.

# Key Word Reference List

The key words that you learned this year are listed here in alphabetical order. If you cannot remember the meaning of a word, turn to the page number that is given next to the word. Read the definition and look at the pictures or examples to help you remember what the word means.

a.m. (page 118)

add (pages 32, 76, 124)

after (page 3)

afternoon (page 118)

area (page 142)

ascending (page 8)

attributes (page 138)

balance scale (page 130)

balanced (page 130)

bar graph (page 93)

before (page 3)

between (page 3, 17)

bill (page 62)

calendar (page 24)

cent (page 60)

centimetre (page 104)

certain (page 150)

change (page 64)

check (page 83)

choose (page 86)

circle (page 70)

classify (pages 72 and 138)

closest to (page 83)

coin (page 60)

cold (page 37)

colder (page 38)

collect (page 91)

compare (page 110)

conclusion (page 91)

concrete (page 91)

cone (page 136)

cool (page 38)

corner (page 70)

count (pages 3, 82, 111, 123)

covered (page 142)

cube (page 136)

cylinder (page 136)

data (page 91)

date (page 24)

day (page 22)

decide (page 86)

decimal point (page 63)

decimetre (page 107)

descending (page 8)

diagonal (page 9)

digits (page 44)

direction (page 55)

divide (page 146)

division (page 146)

dollar (page 63)

double (page 31)

eighth (page 100)

equal (pages 98 and 142)

equal groups (page 146)

estimate (pages 82 and 105)

even (page 18)

evening (page 118)

expand (pages 48 and 76)

face (page 136)

fact family (page 34)

flat shape (page 70)

forwards (page 80)

fourth (page 98)

fraction (page 98)

gram (page 132)

greater than (page 110)

grid (page 56)

groups (page 14)

guess (page 82)

half (pages 98 and 114)

half-past (page 120)

heavier (page 129)

height (page 104)

high (page 40)

horizontal (page 9)

hot (page 37)

hundreds (page 46)

identical (page 114)

impossible (page 150)

kilogram (page 132)

leap year (page 26)

left (page 55)

length (page 104)

less than (pages 15 and 110)

lighter (page 129)

likely (page 150)

line of symmetry (page 114)

list (page 86)

low (page 40)

mass (page 130)